BirdLife International

BIODIVERSITY AND CONSERVATION IN TUMBESIAN ECUADOR AND PERU

Brinley J. Best and Michael Kessler

BirdLife INTERNATIONAL

Wellbrook Court, Girton Road, Cambridge CB3 0NA, U.K.
Tel. +44 (0)1223-277318 Fax +44 (0)1223-277200 E-mail: birdlife@gn.apc.org

BirdLife International is a U.K. registered charity

The designations of geographical data in this book do not imply the expression of any opinion whatsoever on the part of the authors concerning the legal status of any country, territory, or area, or of its authorities, or concerning the delimitation of its frontiers or boundaries.

ISBN 0 946888 26 4

A catalogue record for this book is available from the British Library

Printed and bound in Great Britain by Smallprint (Leeds)

Citation: Best, B. J. and Kessler, M. (1995) *Biodiversity and conservation in Tumbesian Ecuador and Peru*. Cambridge, U.K.: BirdLife International.

Cover illustrations *Front* White-winged Guan *Penelope albipennis,* one of the most threatened Tumbesian endemics (drawing by Michael Kessler). *Back* White-tailed Jay *Cyanocorax mystacalis* (drawing by Richard Thewlis).

CONTENTS

■

TABLES

■

FIGURES

■

BOXES

■

Artwork

Michael Kessler	p. 59, 62, 67, 70, 75, 77, 79, 82, 86, 93, 138, 141, 146, 149, 150, 158.
Richard M. Thewlis	p. 136, 137, 139, 140, 142, 143, 144, 145, 147, 148, 151, 153, 154, 155.
José Cueva	p. 65, 83.
Amanda L. Best	p. 157.

■

Areas of responsibility within the book

Brinley J. Best
Physical and social background, Avifauna, Conservation recommendations.

Michael Kessler
Physical and social background, Biogeography, Vegetation, Conservation recommendations.

The authors may be contacted at the address below:

c/o Americas Division, BirdLife International, Wellbrook Court, Girton Road, Cambridge, CB3 0NA, U.K.

■

ACKNOWLEDGEMENTS

ONE OF the heartening aspects of work in the field of conservation biology is the universal goodwill of colleagues in the field. It is a pleasure to be able to thank all those people who took the time to make available their data, read draft versions of text or help in any manner of other ways.

We are especially grateful to Martin Kelsey who has guided the project throughout and has provided much valuable advice and support. We are also grateful to Adrian Long, Mike Crosby and Alison Stattersfield of BirdLife International, who provided useful information. Special thanks are due to David Wege who has been especially helpful during BJB's visits to BirdLife's Cambridge office and as a regular correspondent; he also helped to produce various maps in the book from BirdLife's database.

The book benefited from the comments of a large number of reviewers who improved its content at the manuscript stage. They are: C. J. Bibby, C. Canady, C. T. Clarke, P. Clarke, S. Clarke, P. Coopmans, M. Crosby, M. Dodson, J. W. Duckworth, J. Fjeldså, P. Greenfield, M. Kelschebach, M. G. Kelsey, N. Krabbe, G. and F. Miehe, A. Long, A. McNab, T. Mulliken, R. Phillips, R. S. Ridgely, M. B. Robbins, C. Rahbek, T. S. Schulenberg, C. Sharpe, J. Tobias, J. Wann, D. Wege and R. S. R. Williams.

The following individuals provided useful additional data for inclusion in the book: P. Coopmans, O. MacBryde, F. Ortiz-Crespo, N. Krabbe, T. Mulliken, M. Pearman, W. Oliver, the late T. A. Parker III, R. Phillips, M. Plenge, C. Rahbek, M. B. Robbins, T. S. Schulenberg, J. Tobias, J. Wann, M. Whittingham and R. S. R. Williams. We are especially grateful to R. S. Ridgely, who provided important field data from his explorations in the Tumbesian region.

Help with MK's botanical work came from B. Ollgaard, J. Jaramillo, P. Moller Jørgensen, M. Rios, K. Romoleroux, C. Ulloa and R. Valencia at the Catholic University, Quito. S. Laegard, K. Lewejohann and G. Wagenitz helped with the handling of the collection.

BJB would like to thank his parents and his wife Amanda, who provided valuable support and helped with the final editing.

■

FOREWORD

Dr Norman Myers

I AM delighted to write a foreword for this important new book.

The Pacific-coast forests of Ecuador are one of the richest biotic sites on earth. During the past few years I have been conducting an analysis of "hot spot" areas, these being localities that feature exceptional concentrations of species with high levels of endemism, and face imminent threat of habitat destruction. I have identified eighteen such areas, and it turns out they contain at least 20% of the Earth's species facing terminal threat in just 0.5% of Earth's land surface. Of the eighteen areas, three are in a league of their own: Madagascar, the Atlantic-coast forest of Brazil, and the Pacific-coast forest of Ecuador. The Ecuador hot spot is outstanding in that it is a meeting point of several distinct areas of endemism; and due to its complex topography and climate systems, it supports several distinct types of forest in the merest sliver of the Earth's land surface. As a measure of its outstanding biodiversity, BirdLife International has documented more than 50 endemic bird species restricted to just a single sector, the so-called Tumbesian centre.

The Ecuador hot spot is unusual in another sense. The forest habitat has been almost eliminated. Less than five percent of the original forest cover remains, and it continues to be depleted. The dry-forest segment is smaller still, yet provides important habitat for many of the hot spots' endemic bird species.

So I warmly welcome this new book. It presents a detailed documentation of a thoroughly imperilled segment of Earth's biodiversity. Fortunately, this is still time for us to turn a profound problem into a magnificent opportunity, provided we can mobilize a conservation effort of a scale to match the challenge. In this tiny locality we can save more endemic bird species, together with associated endemic mammal and plant species, plus untold numbers of endemic invertebrate species, than in much of Europe.

I have been working on the entire problem of mass extinction of species for over a quarter of a century. Depressing as the prospect often seems to me, I am heartened by the many conservation campaigns now underway right round the world. The remarkable effort is epitomized by this new book, symptomatic of the commitment of people in many lands to save our global heritage at stake in all lands. I hope the book enjoys the wide readership it splendidly deserves.

■

PREFACE

Dr Antonio Brack E.
DIRECTOR, UNITED NATIONS DEVELOPMENT PROGRAMME, LIMA

THE PACIFIC coastal region of southern Ecuador and northern Peru, penetrating up to the valley of the Río Marañón, constitutes a zone of great biogeographic interest and complexity.

The forests of the region support high levels of endemism in both flora and fauna, giving the area tremendous importance for the conservation of biological diversity at both national and global levels. Its destruction would carry with it the disappearance of many species, of which about 100 endemic faunal species have so far been described from the few studies carried out. The region also contains important genetic plant resources, especially gourds *Cucurbrita*, chirimoya *Annona cherimolia*, papayas *Carica*, as well as ornamental plants such as *Bougainvilla* and numerous orchids. It contains species of potential commercial use such as the trees *Tabebuia, Cordia* and *Prosopis*, and the fish *Dormitator latifrons*. It also offers the potential for non-destructive commercial activities such as ecotourism.

The countries which share the region have made a laudable effort to establish protected areas, exemplified by such sites as the Machalilla National Park and Manglares-Churute Ecological Reserve in Ecuador and the North-West Peru Biosphere Reserve. However, as financial security for these areas is not guaranteed, their futures are in doubt, and the conservation of further areas is of outstanding importance.

The scientific knowledge of the region is far from complete and it is important to promote and undertake further scientific research. Local universities (Guayaquil and Piura) could develop a programme of investigation and foster an exchange of knowledge; they could also act as centres of monitoring and information.

Finally it is vital that alternative forms of development are established, with the involvement of local people. These should aim to place value on ecosystems in their natural state and promote activities which benefit from an ecosystem without destroying it. These activities could include ecotourism and wildlife management (sport hunting of common species), and the husbandry of native fish and crustacean species. Certainly the responsibility of conserving the region should go hand in glove with the promotion of sustainable activities developed to generate economic alternatives for local people.

I am, therefore, pleased to see the publication of this book as it provides a framework upon which plans can be devised to tackle these important issues before the forests of the region are destroyed.

■

INTRODUCTION

THE BIOLOGICAL diversity of our planet is concentrated in "hotspots" in the tropics where unusually high numbers of endemic species occur (Myers 1988). Such regions were termed "Endemic Bird Areas" by BirdLife International in their publication *Putting biodiversity on the map* (ICBP 1992). They have special conservation importance since they support at least 20% of the world's threatened bird species in a tiny fraction of its surface.

Putting biodiversity on the map provided a badly needed global review of the distribution and conservation of the world's 221 EBAs. That publication highlighted two urgent research priorities dealing with the conservation of specific EBAs:

> " a need to identify the key habitats and sites, assess how well they are covered by the present protected areas network and, where necessary, campaign for increased levels of protection".
>
> "a need to refine the analysis to make more precise recommendations for species-specific actions and to promote the development of conservation action on an adequate scale".

Very few EBAs support more than 50 restricted-range species, making the Tumbesian Western Ecuador and Peru EBA, with 55, internationally significant. Unfortunately since the American Museum of Natural History's early 20th century explorations in the Tumbesian region, most of the area's forests have been destroyed by human activities. The small fragments that remain are highly threatened, yet still support exceptionally high numbers of endemic plant and bird species. Knowledge of the biodiversity of the region, however, remains fragmented in numerous published and unpublished reports and papers, and in the files of biologists who have visited the region. Concern that all available information should be brought together, without delay, into a more accessible and standardized format, prompted us to write this book. It is also a direct response to the call for more detailed research on specific EBAs made by ICBP in 1992 and we hope that it may stimulate others to prepare similar in-depth analyses for additional centres of endemism.

On our visits to the Tumbesian region we have been constantly impressed by the interest and concern of many Ecuadorian and Peruvian nationals towards the wildlife of their country and its conservation. Due to social and economic constraints, however, such concerns inevitably become clouded, and short-term actions designed to sustain people can seriously damage the region's biodiversity. For this reason our book concludes with conservation recommendations designed to meet both the human and wildlife needs of the Tumbesian region. In our view this is the only approach which stands a reasonable chance of success.

As we draw towards the end of the current century, it is becoming increasingly plain that global biodiversity will not survive far into the next century in the state we know it now. Because centres of endemism provide such cost-effective places to concentrate our conservation efforts, they must form the foundations upon which to protect habitats and species. This book attempts to set the conservation ball rolling for just one of the world's 220 or so EBAs.

We will have achieved our aims if this book can provide the impetus for a conservation initiative which can guarantee a sustainable future for the biological diversity of the Tumbesian Centre of Endemism. This diversity includes, of course, our own species, capable of destroying the remaining Tumbesian ecosystems within a generation, but still with the time to secure their long-term survival.

Brinley J. Best and Michael Kessler
May 1995

■

PHYSICAL AND SOCIAL BACKGROUND

GEOGRAPHY AND TOPOGRAPHY

ECUADOR AND Peru are commonly divided into three main regions (Figure 1):

- The western lowlands, which are up to 200 km wide in Ecuador and called the *Región Occidental* or *Costera*, and usually much narrower in Peru where they are simply called *Costa*.
- The Andean chain, running from north to south through both countries, with peaks of over 6,000 m, which geographically as well as climatically separate the coastal plain from the Amazonian lowlands. In Ecuador this region is called the *Región Central* or *Andina*, in Peru it is known as the *Sierra*.
- The Amazonian lowlands, known as the *Región Oriental* (or *Oriente*), *Amazonica* or as *Selva*. A very wet region intersected by numerous rivers which all drain into the Amazon.

These three regions, with differing geological and climatic conditions, have different histories of human settlement and colonization. The Andean valleys witnessed the rising of the Inca culture, one of America's most highly developed, while the lowlands, especially on the eastern side, were until recently only sparsely settled. This book considers only the coastal lowlands and the western Andean slopes and foothills between the Equator and 8° south.

The lowlands themselves may be divided into several topographically distinct sections:

- North of 2°S in Ecuador the lowlands are between 100 and 200 km wide and are flanked on the western side by the Coastal Cordillera which rises to 900 m. Between these hills and the Andean cordillera lies a flat flood-plain up to 80 km wide which gently slopes from north to south and is drained by the rivers Daule and Babahoyo into the Gulf of Guayaquil.

Figure 1. Topography of Ecuador and north-western Peru.
Light stippling marks areas above 500 m, heavy stippling areas above 2,000 m; areas in black are higher than 4,000 m. Small mountain ranges: 1. Cordillera Chongón-Colonche; 2. Chilla Mts; 3. Cordillera Larga; 4. Cerros de Amotape; 5. Cerros de la Brea; 6. Celica Mts; 7. Mts near Sozoranga; 8. Mts above Ayabaca; 9. Mts above Frias.

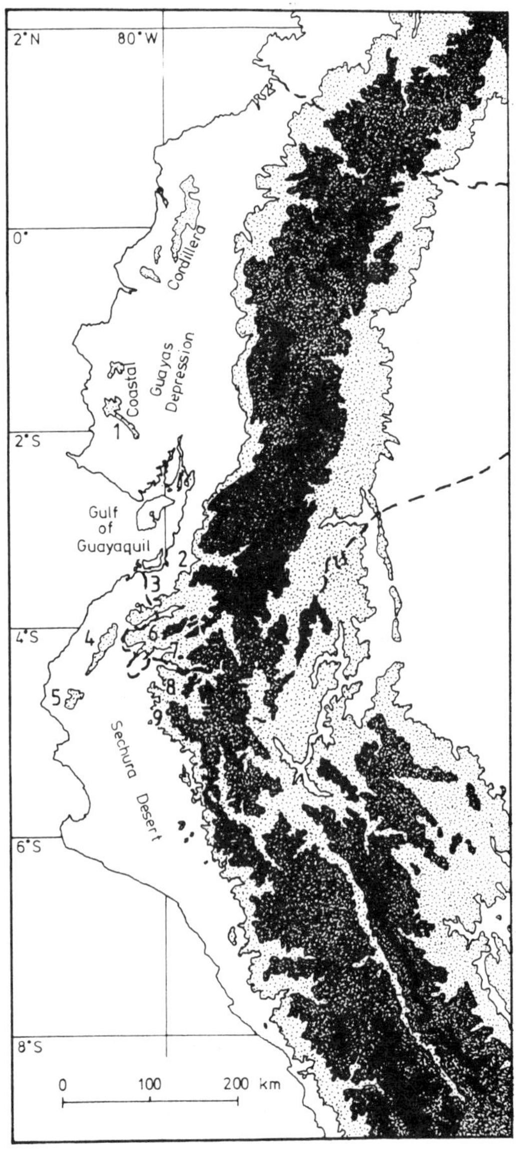

- The Gulf of Guayaquil itself, between about 2°S and 4°30'S, with its numerous inlets and islands, reaches far inland, and reduces the coastal lowlands to a narrow strip no wider than 20-30 km.
- Further south, from 4°30'S to 6°30'S lies another wide area of lowlands, with a hot and arid climate, forming the Sechura Desert. Only a few rivers, mainly the ríos Chira and Piura, intersect this desert and reach the sea.
- South of 6°30'S the Andes reach almost all the way to the coast, restricting the coastal lowlands to a narrow strip which is often no wider than 10 km. This pattern continues, with minor exceptions, to Chile.

Throughout most of Ecuador and northern Peru the Andes form a continuous mountain chain composed of two sections: the Eastern and Western Cordillera. While much of the eastern side slopes off fairly gently, is intersected by numerous rivers and forms a complicated pattern of isolated mountain ranges and foothills, the reverse is true for the western slope. Here, in most areas the horizontal distance from the alluvial plains to the highest mountain tops does not exceed 80 km. The many rivers running down from the mountains only rarely form deep valleys.

At about 4°S, however, this simple pattern is complicated greatly. In this area the general direction of the Andes changes from a north-east to south-west orientation to one running north-west to south-east. The two main Andean chains split up into a number of lower mountain ranges running in different directions (Figure 1). In this region, known as the North Peruvian Low or Huancabamba Depression, no Andean peak reaches over 4,000 m for a length of 400 km. The Porculla pass (2,150 m) in Piura Department, Peru, represents the lowest pass between northern Colombia and southern Chile. For a short stretch, the continental watershed lies on the Eastern and not as usual on the Western Cordillera. The rivers draining west dissect the Western Cordillera into about a dozen small mountain ranges.

These small ranges are named in Figure 1. To illustrate how they are arranged, Figure 2 shows the view from the coast onto the western Andean slope between 3°S and 4°30'S. The mountain ranges become higher from west to east, with only a few, mostly (mountains near Catacocha), or completely (mountains near Cariamanga) hidden. The importance of this arrangement for the distribution of rainfall will be discussed in the climate section below.

For the purposes of this book the Tumbesian Centre of Endemism is defined by the congruent ranges of its endemic bird species (Figure 3). Further information on these can be found in the 'Avifauna' chapter.

Figure 2. View from the west towards the Western Andean Cordillera between 3°S and 4°30′S. Most small mountain ranges in south-western Ecuador and adjacent Peru are shown. Shading indicates relative depth. Vertical axis exaggerated 10 times.

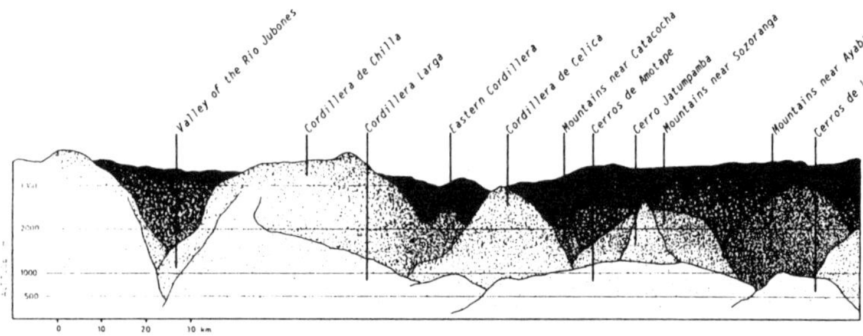

GEOLOGY, SOILS AND CLIMATE

Geology

Over most of western Ecuador and north-western Peru the original Andean metamorphic rocks are covered by Cretaceous, Tertiary and Quaternary volcanic and sedimentary materials. Only locally, e.g. in the area of the Puyango valley, are pre-Cretaceous metamorphic rocks such as slates, serpentines, gabbros, schists, quartzes and gneiss exposed. A few granitic massifs of similar age can be seen: one of the largest is found in Loja, starting at Macará, passing north of Sozoranga and almost reaching Cariamanga.

The upheaval of the Andes started in the late Cretaceous period as the continental South American tectonic plate collided with the oceanic Nazca Plate. Its onset was accompanied by strong volcanic activity, covering most of the Western Cordillera with andesitic lavas, dolerites and pyroclastic material such as ashes, dust and lapilli. Tertiary and Quaternary volcanic activity added rhyolitic and andesitic lavas and further pyroclastic material, leaving few areas free of such rocks.

Erosion began with the uplift of the Andes, leading to the deposition of continental sedimentary detritus, covering depressions and much of the lowlands. Such sandy or clay-rich deposits, conglomerates and breccias can today be found in

Figure 3. The Tumbesian Centre of Endemism as defined in this book.
Based on the ranges of the bird species endemic to the region (see Table 3).

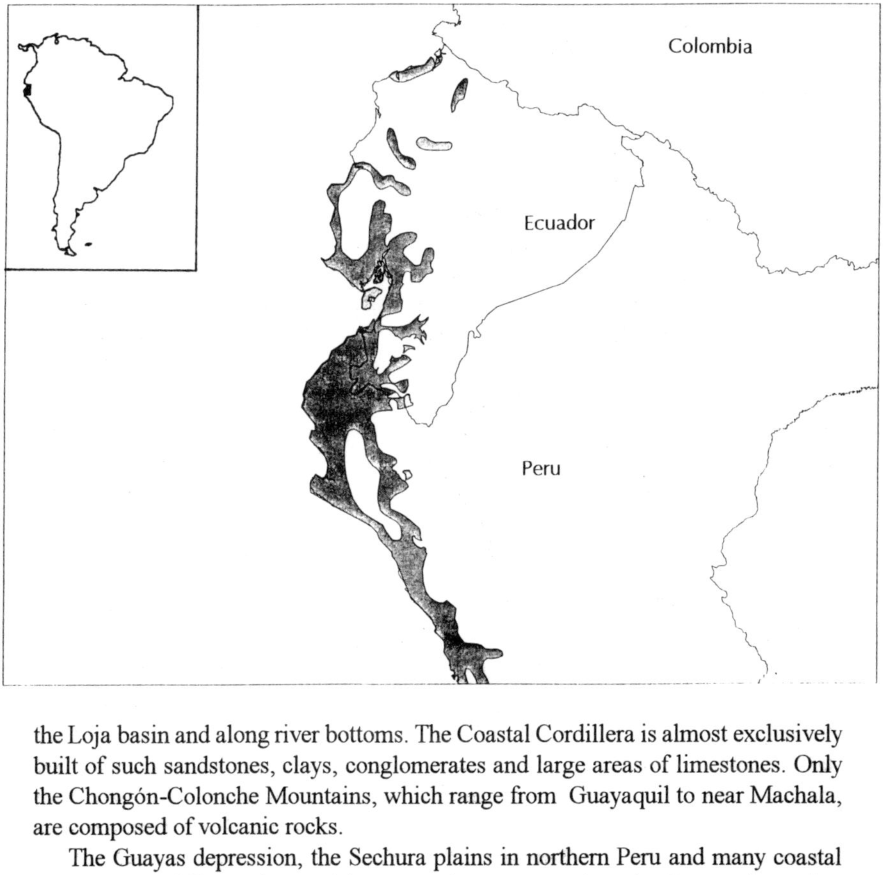

the Loja basin and along river bottoms. The Coastal Cordillera is almost exclusively built of such sandstones, clays, conglomerates and large areas of limestones. Only the Chongón-Colonche Mountains, which range from Guayaquil to near Machala, are composed of volcanic rocks.

The Guayas depression, the Sechura plains in northern Peru and many coastal areas were at different times subject to marine transgressions, leading to alternating periods of oceanic and continental sedimentation. The oceanic deposits included different types of sandy, muddy and calcareous material, while the continental material was mainly alluvial in character.

In summary the Tumbesian region is covered either by volcanic rocks or by sedimentary material, providing a rather uniform nutrient-rich geological source for soil formation. The development of soils is therefore determined mainly by climate, which in the Tumbesian region shows a strong precipitation gradient (see below).

Soils

Desertified areas of northern Peru, and the Santa Elena Peninsula in Ecuador, have weakly developed soils which fall into the categories of arenosols and regosols (soil taxonomy after FAO-Unesco 1988); somewhat more humid areas may have alkaline solonetz soils.

Regions with seasonal climates, i.e. the southern half of the Coastal Cordillera and the Andean foothills from Loja southwards, usually have well-developed, little-leached soils rich in montmorillonite. Such vertisols, planosols, mollisols and luvisols vary in their depth, soil-horizon development, clay content and base saturation, but generally are rather rich in nutrients and are cultivable. Alluvial soils in the Guayas depression and along rivers are also extremely fertile, well developed and have the highest agricultural potential.

Permanently humid areas with abundant rain have soils which are deeply developed, often strongly leached and with high contents of kaolinite. Such acrisols and mollisols are only locally found in the Tumbesian region, mainly along the Andean foothills south to El Oro Province and in the more humid northern part of the Coastal Cordillera. They offer relatively poor opportunities for agriculture. In higher regions with lower temperatures, decomposition slows down, leading to the accumulation of organic material and the formation of the characteristic mollisols of cloud-forests.

Most areas of western Ecuador have a high agricultural potential due to their volcanic or sedimentary origin (Dodson and Gentry 1991). Unlike many tropical regions, western Ecuador can sustain agriculture in the long term in most areas which are not too steep. The region was colonized by pre-Columbian cultures and today is densely settled and intensively cultivated. While the soils of the Andean valleys of south-western Ecuador and north-western Peru could in principle be equally productive, the steepness of the terrain favours erosion and often prevents long-term use; unsuitable agricultural methods have, in fact, already led to intensive erosion and the disappearance of the fertile topsoil in much of that region.

Climate

Because of its geographic position and highly varied topography, the Tumbesian region has a wide variety of climates. The large-scale climatic patterns, dominated by wind systems and sea currents, are modified locally by topographic factors, producing, often over very short distances, conditions ranging from deserts to tropical rainforests. A thorough review of the climate of south-western Ecuador has been made by Munday and Munday (1992); most of the information here is based on their account.

Rainfall is the most variable and therefore most important climatic factor. Its general distribution is determined by the seasonal position of the Intertropical Convergence Zone (ITCZ) which follows the zenithal position of the sun, and by

Figure 4. Approximate distribution of mean annual precipitation in the Tumbesian region. Areas with more than 2,000 mm of rain *per annum* are shaded. Horizontal hatching shows areas which receive considerable additional humidity from mist or fog condensation (relative to local rainfall). Modified and combined after Tosi (1960), Troll (1968), Atlas del Ecuador (1982), Munday and Munday (1992) and data from the Anuarios Meterológicas del Ecuador (1959-1966, 1972-1986) published by the Instituto Nacional de Meterología e Hidrología in Quito.

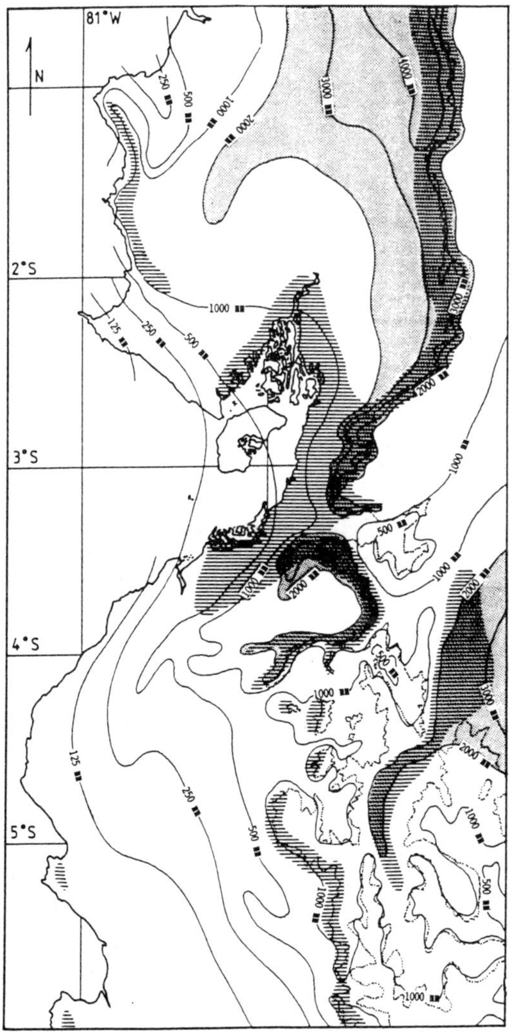

Figure 5. Mean monthly precipitation at a typical site in the Tumbesian region.
Note the pronounced wet season from January to April, the peak month being March. Data from Macará, Loja Province, Ecuador (Altitude 430 m, years 1961-1979), taken from Munday and Munday (1991).

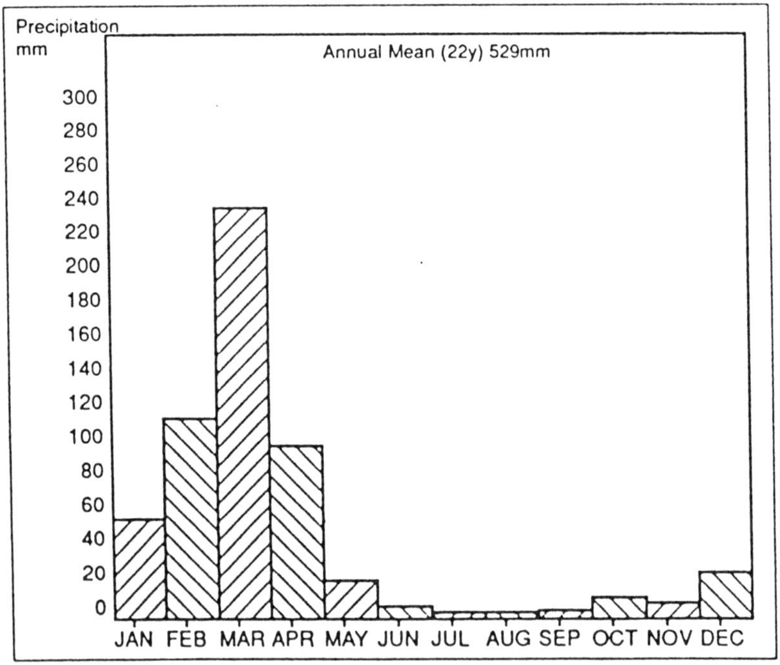

the associated wind systems and sea currents.

From about November to April the dynamic ITCZ reaches its southernmost position along the western South American coast at about 4°S, directly over the Tumbesian region. This brings the associated cloud accumulation and abundant rainfall. The fact that the ITCZ does not reach further south in the Tumbesian region causes the rainfall to decrease sharply from north to south, giving the region its character as a climatic transition zone (Figure 4).

Eastern trade winds over the Amazonian lowlands are of limited importance in the Tumbesian region and only influence the rainfall on the eastern Andean slope (*contra* Munday and Munday 1992). The dominating winds in the Tumbesian region are westerlies, moving inland from the Pacific Ocean (Graf 1986). North of about 4°S along the Ecuadorian coast, they move in over warm sea currents, bringing warm, saturated air which, upon hitting the Coastal Cordillera and Andes rises, cools down and produces abundant orographic rainfall. South of about 5°S the cold Humboldt current flows north along the coast, cooling the air and producing little water vapour. As the winds hit the coast the air heats up, absorbing any available humidity and thereby producing the Atacama Desert which ranges from near Tumbes (3°30'S) well into Chile. In the austral winter, however, cold, saturated air from the sea cools down at night producing abundant fog which gives rise to light drizzling rain (*garúa*) in coastal areas.

There are three major gradients determining the distribution of rainfall:

- Precipitation decreases from north to south as the influence of the ITCZ decreases.
- Westerly winds moving inland shed rainfall on west-exposed mountain ridges. Areas lying further east are located in the rain-shadow and receive less precipitation.
- As humid air moves up along mountains, it cools down and releases rain. This leads, on a continuous mountain slope, to an increase in precipitation with altitude, with a maximum precipitation of between 1,000 and 2,500 mm (usually higher in more arid regions).

The factors mentioned above lead to the complicated distribution of rainfall shown in Figure 4. The highest precipitation levels (over 4,000 mm mean annual precipitation) are to be found along the western Andean slope in the north of the region. Areas with over 2,000 mm reach south to 4°S on the western side of the Celica Mountains; further south precipitation decreases sharply. The Coastal Cordillera also receives more rainfall than the surrounding lowlands, but its low altitude (rarely over 900 m) prevents it from intercepting many clouds and receiving more than 1,500 mm mean annual rainfall.

Rain-shadowed intermontane valleys, especially those of the río Jubones and the río Catamayo in Ecuador and of the río Quiroz in Peru, receive little precipitation, which in the Jubones valley is as low as 250 mm a year. In such valleys daily anabatic (up valley) winds help to dissipate the cloud cover, increasing the insolation, evaporation and aridity of the valleys. Nightly katabatic (down valley) winds, however, may bring cold, humid air from higher elevations, associated condensation and some additional moisture.

The Tumbesian region as a whole shows a single well-marked rainy season from about November to May, with the rainiest month being March in most areas, with between 20% and 45% of the annual rainfall falling during this month (Figure 5). The number of arid months (defined as having less than twice as much monthly precipitation [measured in mm] as the mean monthly temperature [measured in °C]) varies from none in the most humid areas to 12 in the Peruvian Sechura Desert. Throughout most of the Tumbesian region three to nine months are arid.

Annual variability of the rainfall is often extreme. For example, at Zapotillo (mean annual rainfall 695 mm), absolute extremes were 17 mm in 1968 and 3,315 mm in 1983. Exceptionally high amounts of rainfall usually occur in El Niño years (see below). This high variability confronts the local vegetation and wildlife with serious problems, and also complicates agricultural practices in the region.

Besides receiving higher amounts of rainfall, west-exposed mountain slopes are also frequently covered in mist. This has a double effect: it reduces insolation and plant transpiration, and may add considerable amounts of "horizontal precipitation" (see review in Stradtmüller 1987) (Figure 4). While cloud-forests, characterized by such conditions, are commonly found on tropical mountains, along the southern edge of the Tumbesian region a combination of special conditions lead to the occurrence of a very special vegetation type: the *lomas*. Here, in areas which often receive less than 10 mm of rainfall, the *garúas* provide enough humidity to allow tree-growth.

The Gulf of Guayaquil also plays an important role in the distribution of humidity in the Tumbesian region. North of it, the climate is mostly wet; to the south becomes increasingly arid. In the immediate surroundings of the gulf, however, special conditions prevail: the large body of shallow, warm water locally increases temperatures and produces large amounts of warm, saturated air which in turn gives rise to coastal fog and, on the Andean slopes immediately east of the gulf, to markedly increased precipitation levels.

About every 3-16 years the usual pattern of sea currents and rainfall is greatly altered by a phenomenon known as El Niño, or the El Niño-Southern Oscillation (ENSO). In such years, the sea-surface temperatures are much higher than usual because a warm sea-current displaces the cold Humboldt current from the north Peruvian coast and rainfall is much higher than usual (up to 10 times higher than the long-term mean). While the underlying causes are still poorly understood, the catastrophic effects are well known to the local population. In such years exceptional floods and landslides affect agriculture, destroy the infrastructure and cause serious health problems. The effects of the 1983 El Niño event, the strongest ever recorded, are still visible today. Unusually both 1992 and 1993 were El Niño years. In such years, regions which are usually arid become green due to an up-surge in plant growth; it is suspected that bird species which normally frequent only more humid habitat types expand their breeding ranges during El Niño years (e.g. Little Woodstar *Acestrura bombus* [Collar *et al.* 1992]).

Unlike the complicated patterns of rainfall, the temperatures of the region vary very little. In Ecuador, mean annual temperatures are in the order of 22-25°C at sea-level, but decrease along the Peruvian coast with the increasing influence of the cold Humboldt current. Temperature decreases with altitude at a rate of about 0.7°C per 100 m; at 2,000 m the mean annual temperatures are in the order of 10°C. Nightly frost may occur above about 2,400 m. Temperatures change little over the year, these changes are in the order of 1-3°C. Daily temperatures, by contrast, fluctuate by about 10-15°C, greatly exceeding the yearly variation and marking the climate of the Tumbesian region as a typical tropical climate. The range of extreme temperatures (difference between absolute maxima and minima) increases from about 22°C at sea-level to 30°C at 1,200 m and then decreases again to 25°C at 2,500 m. Temperature variation between years lies in the order of 2-5°C, with the hottest being El Niño years.

In summary, it is the geographical and temporal (both inter- and intra-annual) variation in precipitation which is the most important single climatic factor in the Tumbesian region. It determines the distribution of vegetation types and the patterns of distribution in the associated fauna. What remains to be found out, however, are the responses of the plants and animals of the region to exceptionally arid or wet years.

Table 1. Population data relevant to the Tumbesian region.

	Ecuador	Peru
Population (1990)	10, 782, 000	22,332,000
Population projection for year 2000	13,939,000	27,952,000
People per sq. km	40.1	17.4
Urban population	54.2%	69.3%
Rural population	45.8%	30.7%
Population by geopolitical unit (provinces/ departments) [people per sq. km]	Esmeraldas 21.5 Manabí 54.1 Los Ríos 90.5 Guayas 128.7 Azuay 68.0 El Oro 76.1 Loja 36.1	Tumbes 30.4 Piura 41.0 Cajamarca 36.3 Lambayeque 68.1

Source: Brittanica World Data (1991).

POPULATION

Population data on the Tumbesian region are summarized in Table 1. In contrast to Peru, Ecuador is relatively densely settled and also has a larger proportion of its population living in rural areas. The population growth rate of Ecuador (2.4% in 1990) is the third highest of any South American country. The Ecuadorian province of Guayas, which supports the country's largest city, Guayaquil, has the highest population density of any of the provinces within the Tumbesian region, whereas Loja and Esmeraldas Provinces are the least densely settled. Of additional interest is that the province of Loja has a negative migratory balance, mostly to Quito and to Zamora-Chinchipe and El Oro Provinces, whereas El Oro Province itself has a positive migratory balance (Atlas del Ecuador 1982). A possible reason is the lack of surplus agricultural land in Loja Province as a result of past and current intense agricultural activity and land degradation.

PROTECTED AREAS

Both Ecuador and Peru have well developed protected area networks which are essential to the conservation of the biodiversity of the two countries. IUCN (1992) provides a comprehensive review of the legislation and background to conservation in the two countries. The following section deals only with the protected areas.

Categories of protected areas in Ecuador and Peru

The Ecuadorian government officially recognizes 13 categories of protected areas (Box 1) whereas the Peruvian government recognizes eight (Box 2).

The protected areas

There are a number of important protected areas already established within the Tumbesian region, some government owned, others in private ownership (Figure 6). These are described below for Ecuador and Peru (Boxes 3 and 4) and their ornithological importance is detailed in the site directory on pages 162-174. Ecuador and Peru have suffered severe problems in the effective protection of these designated areas. The conservation recommendations chapter discusses these problems on a site specific basis.

Box 1. Protected area categories of Ecuador.

Inter-ministerial Agreement no. 322 (1979)

National Park Areas of 10,000 ha minimum with at least one ecosystem in its natural state and possessing ecological diversity, flora or fauna or geological formations of national, scientific or educational importance. Visitors are only allowed to undertake educational, recreational or investigative activities.

Ecological Reserve Areas of 10,000 ha minimum, with flora or fauna of national importance, particularly endangered species, or geological formations or natural areas of national interest. Natural resources must be maintained in a natural state and exploitation or occupation is prohibited. Educational, recreational and investigative activities only, are allowed.

Faunal Production Area Areas no less than 1,000 ha containing wildlife of commercial value, including traditional subsistence areas of indigenous people. Use of wildlife species is regulated and scientific investigation promoted to allow continuing propagation. Visitors may hunt or collect specimens following established regulations.

National Recreation Area Areas of at least 1,000 ha which are scenically beautiful, have touristic or recreational value and whose ecosystem is in a natural or semi-natural state. They must also be easily accessible to people.

Law of Forestry and the Conservation of Natural Areas and Wildlife no. 74 (1981)

Protected Forest Natural or man-made forested area which meets one of the following criteria: its principle function is soil or wildlife conservation; it is an important watershed area or is adjacent to an important water source; it functions as a wind-break; it is a strategic zone for national defence; forms part of a protected area; or is important for forest research.

Forest Reserve Forest area to remain in its natural state to be used in the future development of the country.

National Park Same definition as under 'National Park' above.

Ecological Reserve Same definition as under 'Ecological Reserve' above.

Wildlife Refuge Areas of any size essential for the preservation of resident and migratory wildlife, or important for scientific, educational or recreational purposes.

Biological Reserve Areas of any size which are in a natural state and are set aside for wildlife conservation.

National Recreation Area Same definition as 'National Recreation Area' above.

Faunal Production Reserve No definition given.

Hunting and Fishing Area Areas where hunting and fishing for sport and food, and controlled recreation is allowed.

Source: IUCN (1992) citing FAO (1982).

Box 2. Protected area categories of Peru.

Forestry and Wildlife Law - Decree Law no. 21147 (1975)

National Forest Forests suitable for timber production, forest products or wildlife, made use of by the state or by individuals with prior authorisation.

Protection Forest Forest of value in soil or water conservation to protect roads, agricultural land and other systems. All exploitation is prohibited.

National Park Areas of wild floral and faunal and scenic beauty where all exploitation is prohibited.

National Reserve Areas set aside for wild faunal species in the interest of the nation. Sustainable harvesting is allowed.

National Sanctuary Areas allocated for the protection of species or communities of plants or animals or any natural formations of scientific or scenic interest.

Historic Sanctuary Protected area where important events in the nation's history took place.

Regulation of the Conservation of Flora and Wildlife, Supreme Decree 15877AG, relating to the 1975 Forestry and Wildlife Law (1977)

Hunting Reserve Privately or publicly owned areas suitable for wildlife management and where the infrastructure allows sport hunting activities.

Communal Reserve Areas reserved for wildlife conservation to benefit local people whose livelihoods traditionally depend on wildlife products.

Reserved Zone Areas designated to protect wildlife until studies are carried out to determine a suitable permanent designation.

Source: IUCN (1992) citing Suárez de Freitas (1990).

Box 3. Protected areas in Tumbesian Ecuador.

Coordinates, altitudinal range and size of each area (if known) are given after the site name.

Cerro Mutiles Reserve, Esmeraldas Province (0°54'N, 79°37'W; 60-300 m; size not known) A small private, moist forest reserve owned by the Luis Vargas Torres (Esmeraldas) University. Although said to be a 'protected' area, recent studies there showed that there is extensive forest logging occurring inside the reserve (Parker and Carr 1992).

Río Palenque Biological Station, Pichincha Province (0°30S, 79°30'W; 200 m; 87 ha of forest) A very small area of moist forest owned by the University of Miami.

Jauneche Biological Station, Los Ríos Province (1°20'S, 79°35'W; 50-70 m; 138 ha) A small moist forest owned by the National University of Esmeraldas.

cont.

Box 3 (cont.)

Machalilla National Park, Manabí Province (1°21'S-1°40S, 80°38'W-80°50'W; 0-700 m; 55,000 ha) Estimates of the proportion of this National Park which is government owned vary from 80% (IUCN 1982) to a mere 25% (Arriaga 1987); it comprises a marine and a continental section with fairly extensive tracts of deciduous, semi-evergreen and evergreen forest. The park, established in 1979, is officially split into five zones: primitive, primitive-scientific, extensive use, intensive use and special use. It is the most important protected area in Tumbesian Ecuador.

Cerro Blanco Reserve, Guayas Province (2°10'S, 80°02'W; 100-420 m; 2,000 ha) A small area of semi-evergreen forest on a ridge behind an Ecuadorian National Cement Company plant. Now managed by Fundacíon Natura and Fundacíon Pro-Bosque.

Manglares-Churute Ecological Reserve, Guayas Province (2°19'S-2°36'S, 79°34'W-79°49'W; 0-900 m; 35,000 ha) 98% of this Ecological Reserve, created in 1979, is government owned (IUCN 1982); it comprises a mangrove and salt-flat area and a forested section.

In addition to the above areas there are three Protected Forests in the Tumbesian region: **Chongón** [Guayas Province; 2,000 ha], **Molleturo** [Azuay Province; 28,100 ha] and **Puyango Petrified Forest** [2,658 ha]. However the establishment of these areas has not prevented the destruction of forest within them. The **Arenillas Military Reserve** in El Oro Province is a further 'protected area', but virtually nothing is known about the reserve, the only reference being Parker and Carr (1992).

Box 4. Protected Areas in Tumbesian Peru.
Coordinates, altitudinal range and size of each area (if known) are given after the site name.

North-West Peru Biosphere Reserve, Tumbes Department (0-1,600 m; 226,300 ha) A large area of deciduous and semi-evergreen forest comprising the larger part of three separate Peruvian government owned protected areas: Tumbes National Forest [3°40'S-3°46'S, 80°16'W-80°21'W; 75,102 ha; created in 1957], Cerros de Amotape National Park [c.80°37'S, 4°05'S; 200-1,600 m; 91,300 ha, created in 1975] and El Angolo Hunting Reserve [65,000 ha; created in 1975]. The area as a whole is the most important forest in the Tumbesian region.

Quebrada Negrohuasi Reserve, Lambayeque Department (6°21'S, 79°29'W) A tiny private reserve set up for *Penelope albipennis*; protection for the reserve apparently exists on paper only (Collar *et al.* 1992).

Figure 6. Protected areas in the Tumbesian region.

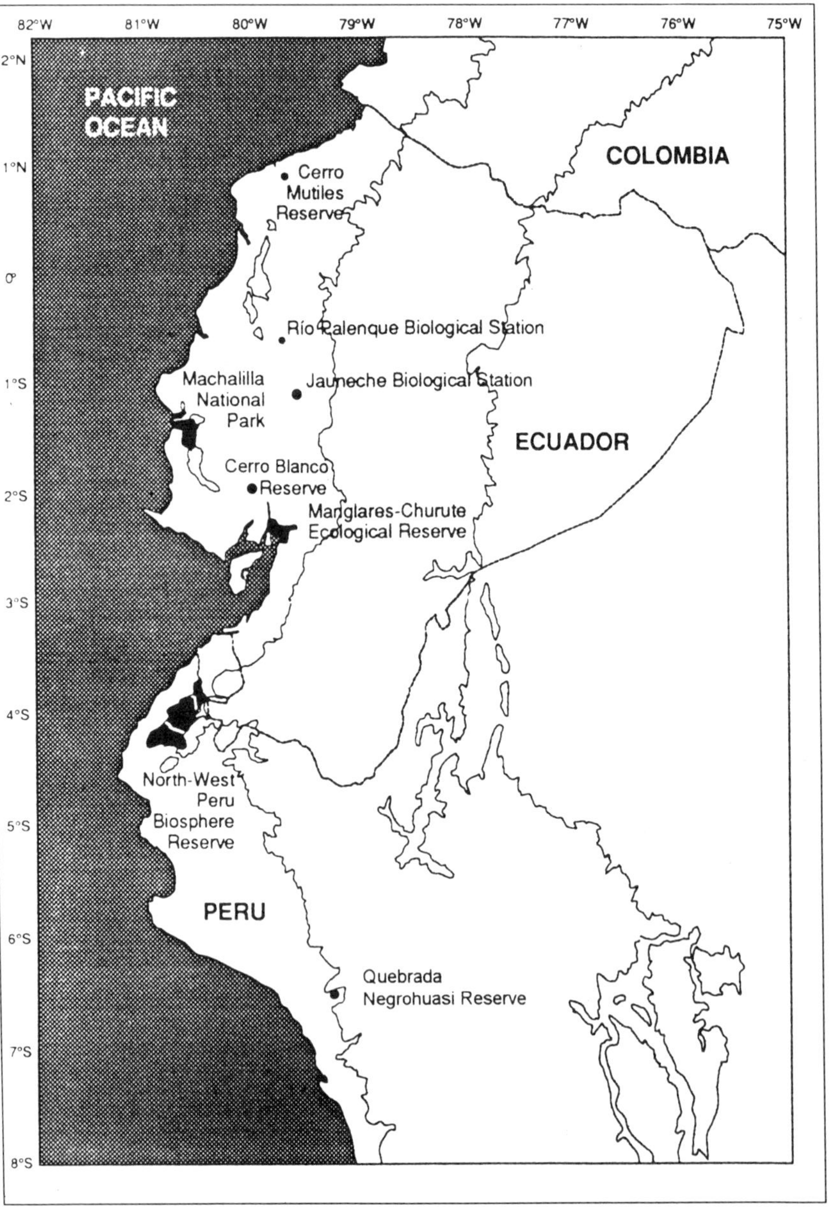

THE HISTORY OF FOREST DESTRUCTION IN THE TUMBESIAN REGION

To those who have travelled through the Andean foothills and western lowlands of Ecuador and adjacent north-western Peru, it will not come as a surprise to learn that less than 5% of the land surface is currently forested. The landscape is a patchwork of crop-land, only occasionally punctuated by small forest patches, themselves often confined to steep, uncultivable slopes. Only at a handful of sites does forest stretch unbroken for a few kilometres of more. The forest that remains is often degraded, especially in the understorey.

Since the mid-1980s several studies have highlighted the region as one of the most threatened on earth in terms of biological extinction as a result of human activities (e.g. Simberloff 1986, Myers 1988), yet not until the last few years were quantitative estimates made of the extent of deforestation, based on satellite images and field reconnaissance. Data are still very scanty from Peru, precluding any detailed discussion of forest destruction there, but an important paper by Dodson and Gentry (1991) concerning western Ecuador focused the world's attention on the plight of the Tumbesian forests and their wildlife.

Dodson and Gentry (1991) present an interesting historical record of forest destruction in western Ecuador (defined as the region from the coast to the 900 m contour line of the Andes, thereby including some non-Tumbesian areas), which, supplemented by additional material, is summarized below.

- Prior to the Spanish invasion of the early 1500s, western Ecuador supported a substantial population of indigenous people. The rural population may have been larger than today's, but as these early people had only waterways and forest-trails as transport routes, their influence on the forest cover was limited. Before their population was drastically reduced by European diseases, they probably accounted for a small loss in forest cover.
- After the decline of these earlier inhabitants much forest regeneration occurred and subsequent modification was localized to the edges of water-courses and along mule-trails. Historical accounts from the 1860s from the Guayaquil region (e.g. Hassaurek 1967), indicate that the region may at that time still have been largely surrounded by unbroken forest.
- The comparatively fertile soils of much of western Ecuador allowed a large sustainable agricultural system to develop. The vast, productive banana plantations of the south-western lowlands, which make Ecuador the world's largest exporter of bananas, demonstrate the fertility of the region's soils.
- Substantial deforestation did not began after 1945 when road construction was accelerated. A tentative map of relatively undisturbed forest in western Ecuador in 1938 (Figure 7a), derived from data on the 1938 road network

Figure 7. Forest cover in western Ecuador below 900 m, 1938-1988.
a: 1938, b: 1958, c: 1988. Note the marked forest decline since 1958 (see text for further comments). Redrawn from Dodson and Gentry (1991).

and rivers navigable by launch, shows that approximately 75% of the surface may then have been covered by forest.

- The period from 1958-1988 was the most destructive in terms of forest loss (from 63% to less than 8% forest cover, Figure 7b and 7c). A number of factors played important roles in this dramatic loss:

  -the population of Ecuador rose from 4 to 10.2 million;
  -land reforms in the early 1960s promoted colonization of government-owned land;
  - hacienda owners were reluctant to keep land in an 'unproductive' state (i.e. forested) through fear of clearance and occupation by landless farmers;
  -the Ecuadorian petroleum industry developed to provide more than 60% of the national income; funds were injected into road-building to provide communication between markets and cities. An extensive network of primary and secondary roads was developed throughout most of western Ecuador, increasing threefold the size of the road network between 1957 and 1988;
  - migration of people into the western lowlands increased the pressure on the newly opened up forests;
  -bananas, oil palm, soya-beans, rice and corn as well as the more traditional cacao and coffee became important export crops, supplying significant agro-industrial income;
  - penetration roads were constructed, largely between 1965 and 1975; these were intended to help put into production all 'unproductive' land, including the forests of western Ecuador. These now form a dense network which allows easy access to most forested areas in the dry season;
  - desertification of parts of coastal Guayas, El Oro and Loja Provinces in Ecuador and Tumbes and Piura Departments in Peru has occurred, caused partly by inappropriate land-use techniques. It has accounted for a 31.5% increase in arid land in the 25 years prior to 1988 (IUCN 1988) and has caused some desert encroachment into former areas of dry forest;
  - expansion of the shrimp-farming industry led to the destruction of part of the mangrove forests bordering the Pacific Ocean; they were reduced by over 10% up to 1984.

THE CURRENT FOREST COVER OF THE TUMBESIAN REGION

The above factors have combined to leave western Ecuador largely deforested: Dodson and Gentry (1991) estimate that only 4.4% of the land surface is now covered by forest (8% of the 1938 cover). Figures 8-11 show the remaining forest cover of four broad types: dry, moist, wet and pluvial.

All except the pluvial forests have suffered catastrophic forest losses since the 1950s, the pluvial forests suffering least due to their poorer soils, their inaccessibility and the broken nature of their terrain. There are only two areas of extensive forest remaining, both of which have designated reserve status: the Awá Reserve (80,000 ha, with 120,000 ha of uncommitted forest around it), and the Cotacache-Cayapas Ecological Reserve (204,420 ha), these being in the provinces of Esmeraldas, Carchi and Imbabura in north-western Ecuador. Although these areas have extreme biological importance and may support the most diverse and biologically unique forests on earth, biologically they belong to the Chocó Centre of Endemism and therefore will not be discussed in this book.

The largest forest fragments left in the Ecuadorian sector of the Tumbesian region lie in west-central Ecuador in the Machalilla National Park. In north-west Peru is the largest continuous tract of forest in the entire region: the North-West Peru Biosphere Reserve (75,102 ha; see page 33). Together the Machalilla and Tumbes reserves are vital to the continued survival of many of the habitats and species distinctive to the centre of endemism. The importance of the Mache-Chindul Mountains which lie at the northern end of the Coastal Cordillera and apparently contain extensive tracts of undisturbed forest (Mudd 1991), has yet to be established. They are situated on the very northernmost edge of the centre of endemism, beyond the known range of several of the Tumbesian endemics.

Figure 8. Dry forest cover in western Ecuador below 900 m in 1958 (left) and 1988 (right). (Redrawn from Dodson and Gentry 1991).

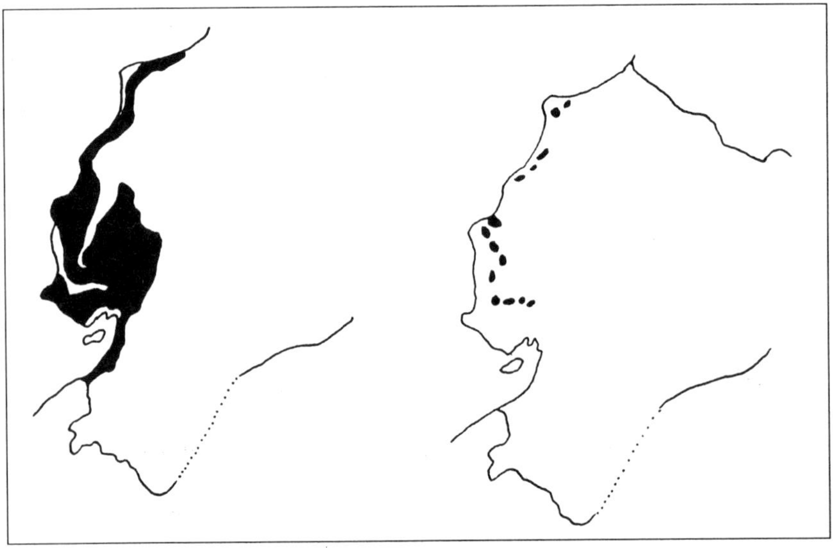

Figure 9. Moist forest cover in western Ecuador below 900 m in 1958 (left) and 1988 (right). (Redrawn from Dodson and Gentry 1991).

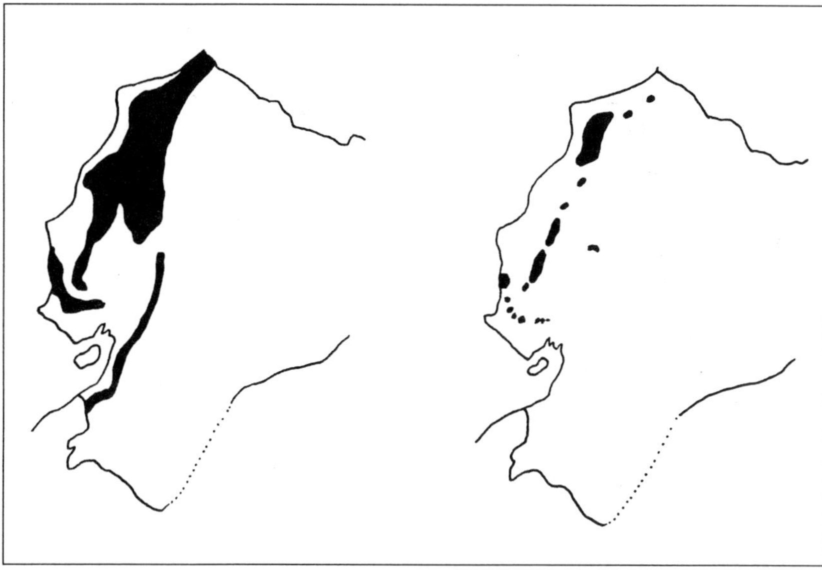

Figure 10. Wet forest cover in western Ecuador below 900 m in 1958 (left) and 1988 (right). (Redrawn from Dodson and Gentry 1991).

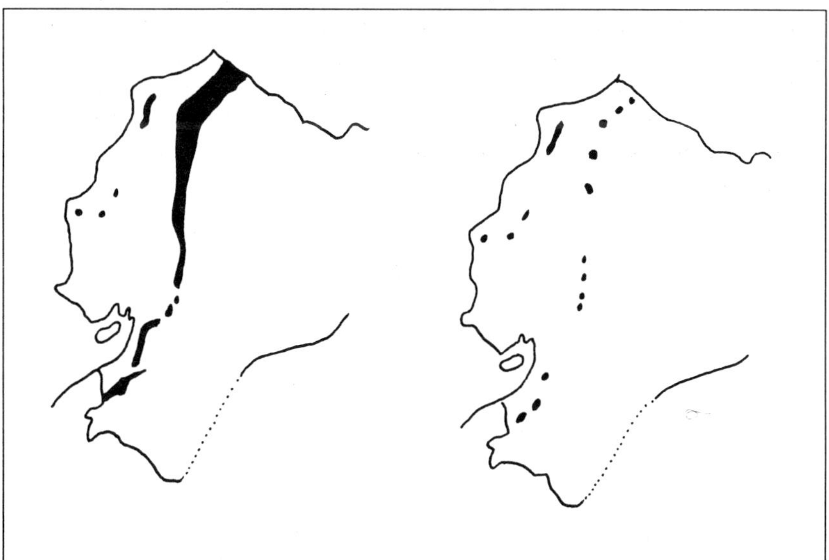

Figure 11. Pluvial forest cover in western Ecuador below 900 m in 1938 (left) and 1988 (right). (Redrawn from Dodson and Gentry 1991).

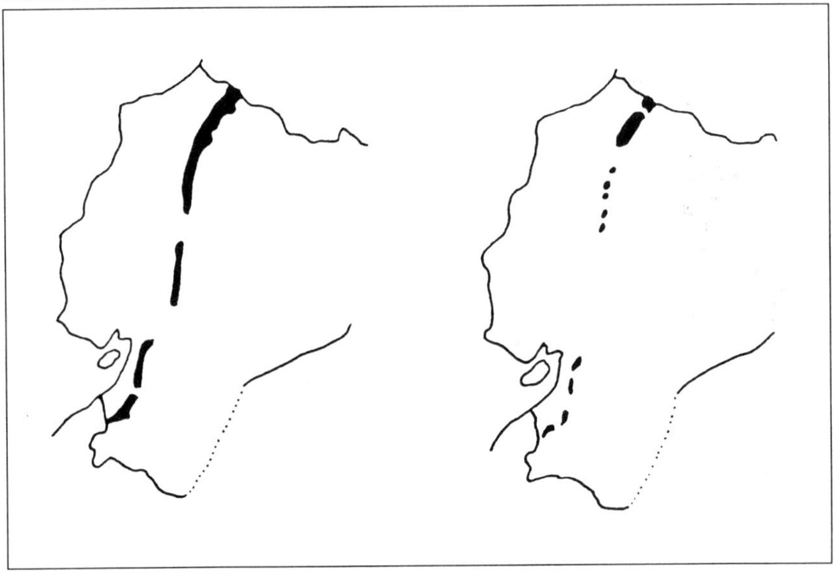

Outside the North-West Peru Biosphere Reserve there are very few areas of forest in north-western Peru greater than 1,000 ha in size. Small forest patches (between 100 ha and 3,000 ha in size) still occur in lowland western Ecuador. These are few in number and are mostly part of the privately-owned reserves detailed on pages 32-33. Over the major part of the Tumbesian region forest is confined to tiny, isolated and fragmented patches less than 100 ha in size, which cling on to the steepest slopes where agriculture is most difficult. These last areas are now at risk from agricultural expansion, selective logging and grazing cattle which trample and browse the understorey.

■

BIOGEOGRAPHY

INTRODUCTION

BIODIVERSITY IS not evenly distributed on earth. Some areas stand out as containing particularly large proportions of localized species, whereas others support very few endemics. The Tumbesian region stands out because it supports one of the highest numbers of restricted-range bird species of any Endemic Bird Area (Stattersfield *et al.* in prep.). This chapter analyzes whether this endemicity is also shown by other groups of organisms and discusses a series of historical and ecological hypotheses for the development of this centre of endemism. The general paucity of knowledge on the species composition and biogeography of the region makes these discussions rather speculative.

IS THE TUMBESIAN CENTRE OF ENDEMISM UNUSUALLY RICH IN RESTRICTED-RANGE SPECIES?

Based on the range limit of 50,000 km^2, Terborgh and Winter (1983) and ICBP (1992) found that the Tumbesian Centre of Endemism contains one of the highest concentrations of restricted-range bird species in South America. However, if Cracraft's (1985) concept (that a centre of endemism should, regardless of size, be defined by the congruence of the range limits of its constituent species) is applied to the six dry forest areas recognized in the Neotropics by Gentry (1992) (Figure 12), it emerges that the Tumbesian Centre of Endemism does not stand out (Table 2); it is merely because it is small enough to fulfil the 50,000 km^2 criterion that suggests that it is particularly rich in endemic species. In this discussion 'dry forests' include lowland forests with up to 1,600 mm annual precipitation, thereby including the majority of vegetation types of the Tumbesian region.

A similar conclusion was derived by Gentry (1992) for the floras of the six dry forest areas. While the incomplete database allows only tentative conclusions, it shows that the western Mexican dry forests have the highest concentrations of both endemic genera and species; the Tumbesian dry forests seem to show an average level of endemism. Dodson and Gentry (1991) estimated the proportion of plant species endemic to lowland western Ecuador and adjacent areas of Colombia and

Figure 12. Distribution of dry forests in the Neotropics.
Slightly modified after Gentry (1992) and Hueck and Seibert (1981). Following Gentry (1982, 1992) the areas are arranged into six groups: Antillean (including Yucatán), Western Mexican, Central American, Northern South American, Coastal Ecuadorian/Peruvian (Tumbesian)[arrowed] and South-eastern South American.

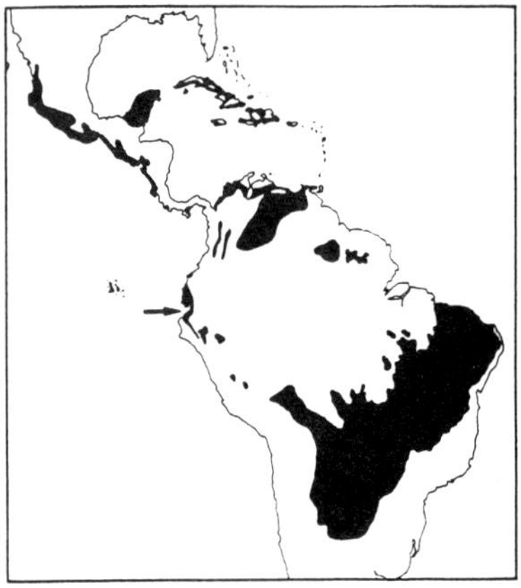

Table 2. Bird endemism in six Neotropical dry forest regions.
Approximate number of bird species endemic to the six dry forest areas of Gentry (1992).

Dry forest area	Number of endemic bird species
Tumbesian	55
Western Mexican and Central American	54
Antillean dry forest specialists	35
also in other habitats	44
total	79
Northern South American	25
South-eastern South American	
Caatinga	20
Cerrado	40
Chaco	36
shared between two or three	26
total	122

Sources: Meyer de Schauensee (1970), Peterson and Chalif (1973), Bond (1980), Vaurie (1980), AOU (1983, 1987, 1989), Cracraft (1985), Ridgely and Tudor (1989), Best and Clarke (1991), Stattersfield *et al.* (in prep.) and personal field experience.

Peru to be 20% of the total flora, or 1,260 of 6,300 species. They stressed that many endemics are restricted to narrow ecological niches within very small ranges. Many species are restricted to isolated habitat pockets, e.g. mountain ranges. Brown (1982) found no area of butterfly endemism directly equivalent to the Tumbesian centre.

Thus, the subjective limit of 50,000 km^2 for defining endemic bird species will tend to over-emphasize the importance of areas which are intrinsically small, e.g. oceanic or habitat islands. For biogeographical analyses only the inclusion of a principle like the one employed by Cracraft (1985) can compensate for the artificial boundaries imposed by setting an upper geographical limit for the range of endemic species. For conservation purposes, however, the 50,000 km^2 threshold is useful since it enables areas of conservation priority to be found.

Although the Tumbesian centre of endemism is not significantly richer in endemic bird and plant species than other dry forest areas, Ecuadorian dry forests have more endemic plant species (20% of the local flora from Capeira [Dodson and Gentry [in press]) than the adjacent moist forest at Jauneche (15%, Dodson and Gentry 1985) and the moist forest on Barro Colorado Island, Panama (12%, Croat and Busey 1975). A similar conclusion was reached by Rzedowski (1978) in Mexico where 43% of the endemic genera are from arid areas, while only 28% are found in semi-arid areas, 11% in semi-humid areas and 4% in humid areas. This is also true for birds, with over 50 species restricted to the Tumbesian region but only two or three to the rest of lowland western Ecuador.

This leads us to wonder how the comparatively high degree of endemism of the dry forest has developed. Traditionally there has been a dispute between authors invoking mostly historical factors (e.g. Haffer 1969, 1974, 1982, Müller 1973, Prance 1982, Whitmore and Prance 1987) and those considering that current ecological relationships are more important (e.g. Nelson and Platnik 1981, Rosen 1978, Vuilleumeir and Simberloff 1980).

HISTORICAL FACTORS

By the mid-Miocene (about 14 million years ago) the Andes had risen to such a height that topographic and climatic conditions similar to today's had developed. Along the Pacific coast of Peru the tropical forests, which originally covered much of tropical South America, were replaced by the desert which is still present today (Simpson 1983). Until about 25 years ago, it was generally believed that since the establishment of these conditions, tropical areas had suffered little climatic disturbance, especially during the era of Pleistocene glaciations (2 million years ago to 10,000 years before present [BP]). However, since then extensive palaeobotanical and palaeoclimatological evidence has shown that the tropics were subject to severe climatic changes during the last ice-age (e.g. Wijmstra and van der Hammen 1966, Hooghiemstra 1984, Ochsenius 1984, Tricard 1985 and many others) leading to the development of the theory of Pleistocene forest refugia (e.g. Haffer 1969, Prance 1982, Whitmore and Prance 1987).

During drier (and cooler) Pleistocene times, humid forests were restricted to topographically and climatically favourable sites, leading to speciation among then isolated populations of forest-inhabiting species, though there is a lack of supporting palaeoclimatological and palaeogeological evidence for this. It now seems likely that the areas most affected by climatic changes were the transition habitats between wet and arid vegetation types (Kutzbach and Guetter 1984, 1986, Colinvaux 1987). Here, even slight climatic modifications would have led to considerable disturbances of the plant communities, while very dry or humid areas would have been less affected by a decline or rise in precipitation levels. The Tumbesian region represents such a transition zone, lying between the very wet rainforests of north-western Ecuador and western Colombia and the very dry coastal deserts of Peru and Chile.

How would the Pleistocene climatic changes have affected the Tumbesian region? Among Neotropical dry forest areas, the Tumbesian region is unique in lying close to the equatorial latitudes. Unlike the other dry forest areas whose presence is caused by their position within the arid subtropical climate belts (southern and eastern South America, Central America), by edaphic factors and fire (the llanos of Venezuela, the Guianan savannas) or by their position in rain-shadowed valleys (Andean intermontane valleys), the climate of the Tumbesian region is determined by the 'collision' of the cold Humboldt (or Peruvian) sea current from the south with warm currents from the north and west. By its general cooling effect and its limited evaporation, the cold Humboldt current gives rise to an extremely arid climate along the Chilean and Peruvian coasts. Warm sea currents, on the other hand, produce warm, saturated air which, upon hitting the continental land-mass, give rise to abundant rainfall. What is of primary importance for an analysis of the Pleistocene history of the Tumbesian region, is that the position of the contact zone is determined by the Equator. Between these two sea currents the warm equatorial counter-current moves in from the west, bringing warm water to areas somewhat south of the Equator.

For the last ice-age the general precipitation decline at tropical latitudes in South America has been estimated to have been between 20% (Kutzbach and Guetter 1984, 1986) and 50% (Leyden 1985), while subtropical latitudes (20° to 38°S) were effectively moister (Markgraf 1989). It seems likely that during the last glacial maximum (about 18,000 BP) the sea surface temperature was lowered by about 4-6°C (Rind and Peteet 1985, Colinvaux 1987, Lauer 1988), while during the post-glacial hypsithermal (about 8,000-6,000 BP) there was a climatic maximum with temperatures about 2°C higher than today's (van der Hammen 1981, Lauer 1988). However, this probably did not alter the basic pattern of sea currents

(Campbell 1982). Therefore, during the last ice-age the relative effect of the Humboldt current was strengthened, probably leading to a more pronounced aridity in the southern portion of the Tumbesian region, while during the post-glacial hypsithermal the warm currents reached somewhat further south, probably leading to conditions similar to those during the present-day El Niño years when warm water-masses reach further south along the Peruvian coast than usual, bringing exceptional amounts of rain. But these changes, regardless of their impact, did not lead to a considerable shift in the position of the vegetation types of the Tumbesian region, because its location at the transition zone between the wet north and the dry south has a fixed position at all times, which was determined by its position close to the Equator.

It seems probable that the wet forests of western Colombia and north-western Ecuador (the Chocó region) were restricted in extent during the Pleistocene glaciations. This is suggested by a close floristic similarity at a generic level between the Tumbesian dry forests and the dry plant formations of the Caribbean coasts of Colombia and Venezuela (Sarmiento 1975, Campbell 1982, Gentry 1992), although this might also be explained by present-day connections through the arid inter-Andean valleys of the northern Andes. However, range disjunctions between the dry forests of Central America and those of the Tumbesian region, as exemplified by the White-winged Dove *Zenaida asiatica* and the closely related allospecies Scrub Blackbird *Dives warszewiczii* and Melodious Blackbird *Dives dives*, indeed indicate that these dry forest areas must have been connected in recent times.

The large number of endemic plant and bird species found in the wet Chocó forests indicate that some wet forests persisted through the ice-age. In fact, even a precipitation depression of 50% would have left extensive evergreen forest areas in a region which today receives up to about 10,000 mm annual precipitation. Even in the Tumbesian region, some areas receive over 3,000 mm annual precipitation. It seems probable that during the Pleistocene, dry forests extended further north along the coast, confining wet forests to the Andean foothills and coastal hills. This increased the range of the dry habitats of the Tumbesian region, but did not displace them entirely from their present positions.

However, Campbell (1982) based on analysis of fossil remains from Pleistocene deposits in north-western Peru and south-western Ecuador, concluded that during the last glacial period the Tumbesian region was more humid than today. This different scenario is based on a different interpretation of possible changes in the distribution of trade winds, which play a crucial role in the transport of air masses from the sea to the mainland. In accordance with other models, a more arid climate is postulated for the Colombian coast, leading to a dry forest corridor connecting the Tumbesian area with the dry forests in Central America.

The final important factor to be considered is how the Pleistocene temperature depression would have affected the Tumbesian region. Given our current state of knowledge, a temperature decrease of 5-10°C seems most probable (Flenley 1979, Hooghiemstra 1984, Lauer 1988, Markgraf 1989, Bush *et al.* 1990), leading to an altitudinal lowering of vegetation zones by about 1,000 to 2,000 m, assuming the same temperature lapse rates as today. Any vegetation-type dependent on high temperatures (i.e. the lowland conditions) would have been displaced. It is interesting to note in this context that the highest level of plant endemism occurs in Ecuador at intermediate altitudes between 1,000 and 3,000 m (Balslev 1988), i.e. in those areas whose climatic conditions existed throughout the ice-ages, even though at different altitudes than today. But what happened in the Tumbesian region? Most vegetation types reach altitudes of over 1,000 m, especially in the drier southern part of the region (see 'Vegetation' chapter). Only a few vegetation types (e.g. the moist lowland forest) are confined to low elevations, but this does not mean that the component species would not be able to survive periods with lower temperatures.

In summary, the Tumbesian dry forests probably persisted throughout the Pleistocene at roughly the same location as today (Figure 13), therefore representing a glacial refuge (though of a different type than the Pleistocene wet forest refugia of Haffer [1969], Prance [1982] and Whitmore and Prance [1987]).

Figure 13. Schematic distribution of vegetation zones on the western Andean slope between 6°S (south-western Colombia) and 14°S (central Peru) at present (A), during the height of the last glacial period (B) and during the post-glacial climatic optimum (C). Modified after Lauer (1988). The hatching shows the approximate present-day distribution of the Tumbesian region. Note how the location of the most important Tumbesian vegetation zones remained relatively stable during periods of climatic change.

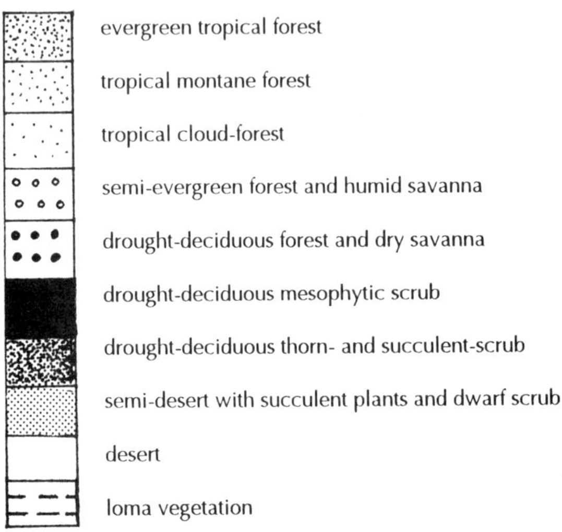

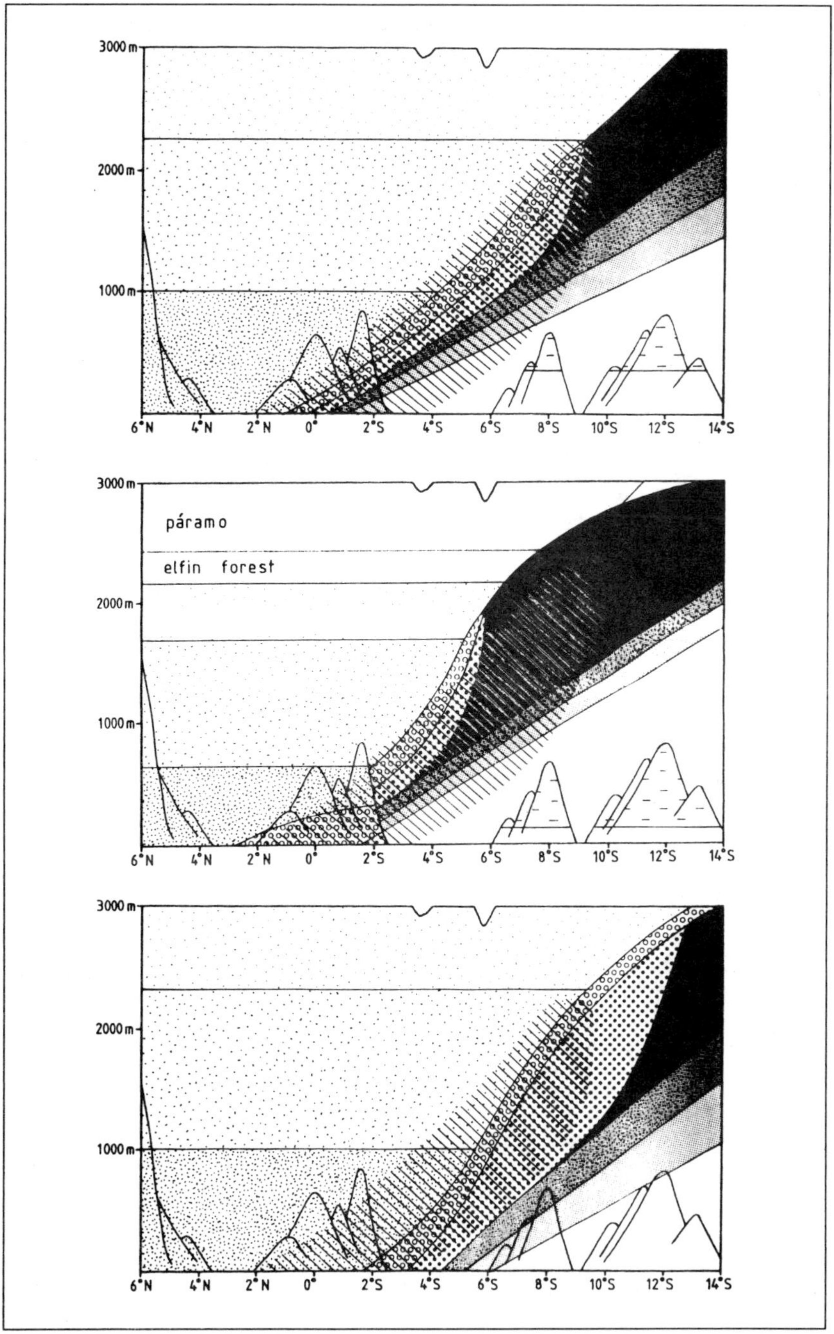
3000 m
2000 m
1000 m
6°N
4°N
2°N
0°
2°S
4°S
6°S
8°S
10°S
12°S
14°S
páramo
elfin forest

SPECIATION IN THE TUMBESIAN CENTRE OF ENDEMISM

Different groups of organisms may react in dissimilar ways to historical and ecological factors. Although the dry forests of western Mexico have in Gentry's (1992) analysis by far the highest degree of generic plant endemism, when bird endemism is considered, this area is not outstanding (Table 2). As the following analysis is based on birds (the only group known well enough for a detailed study), the results should be interpreted with care.

Table 3 divides Tumbesian habitats into five broad categories (arid thorn-scrub to humid cloud-forest). The endemic bird species occur in all vegetation types, not just in the dry forest. Fifteen (27%) of the endemics listed in Table 3 occur only in a single habitat type, with 29 (52%) confined to just two. Eleven species (20%) are found in three habitat types, only one in four types, and none in five. The habitat with the highest number of unique species is arid scrub and woodland (12 species, 21%), with 10 species (18%) occurring only in this habitat category and dry deciduous forest. Thirteen species (22%) are confined to deciduous and semi-evergreen forests and eight (14%) to semi-evergreen and evergreen forests.

The resulting question is whether the endemic species originated through isolation of this patch of dry forest that was connected to other dry forests in Pleistocene times (see above), or whether the species colonized this isolated pocket from surrounding habitats (wet forests, Andean habitats), adapted to a different habitat and reached reproductive isolation. Of the 35 Tumbesian bird species whose presumed closest relatives are known (Table 3) only 15 (43%) of them have their closest relatives in other arid regions (including the isolated Marañón valley which was as likely colonized from the Tumbesian region as *vice versa*). Seven species' (20%) closest relatives are from Andean habitats, five (14%) from humid Amazonian lowlands and three (5%) from humid lowlands of north-western South America. Hence, the Tumbesian avifauna is not merely part of a generalized Neotropical dry forest fauna. This compares well with Gentry's (1992) results that at the generic level the flora of the Neotropical dry forests is essentially an impoverished version of that of the moist and wet forests, and lacks a substantial portion of its own floristic elements. A different conclusion was arrived at by Chapman (1926) and Marchant (1958) who emphasized the avifaunal connections between the Tumbesian region, the arid Marañón valley and the dry forests of Central America.

However, in spite of these obvious connections with other dry forest areas, it seems likely that a considerable proportion of the endemic bird species originated locally from species invading the Tumbesian region from adjacent (e.g. humid lowlands of north-western South America, humid Andes) or distant (e.g. lowlands of eastern South America) sources. While the latter case is easily explained by allopatric speciation, this is less likely for species whose closest relatives are at present only narrowly allopatric, parapatric or even sympatric. Possible

Table 3. Restricted-range bird species of the Tumbesian Centre of Endemism, with habitat preferences and presumed closest relatives.

English name	Scientific name	Habitat	Presumed closest relative	Distribution of relative
Pale-browed Tinamou (PBT)	*Crypturellus transfasciatus*	D, S	?	-
Grey-backed Hawk (GBH)	*Leucopternis occidentalis*	S, E	*L. albicollis*	widespread, lowlands and foothills
Rufous-headed Chachalaca (RHC)	*Ortalis erythroptera*	S, E	*O. garrula?*	humid N lowlands
White-winged Guan (WWG)	*Penelope albipennis*	D	*P. purpurascens?*	humid N lowlands
Ecuadorian Ground-Dove	*Columbina buckleyi*	S	*C. talpocoti*	widespread, humid lowlands
Ochre-bellied Dove (OBD)	*Leptotila ochraceiventris*	D, S, E, W	*L. cassini?*	humid N lowlands
Red-masked Parakeet (RMP)	*Aratinga erythrogenys*	D, S, E	*A. wagleri/A. mitrada*	Andes
El Oro Parakeet (EOP)	*Pyrrhura orcesi*	W	*P. melanura*	humid E lowlands
Pacific Parrotlet	*Forpus coelestis*	A, D, S	*F. xanthops*	Marañón endemic, dry forest
Grey-cheeked Parakeet (GCP)	*Brotogeris pyrrhopterus*	D, S	?	-
Scrub Nightjar	*Camprimulgus anthonyi*	A, D	?	-
Tumbes Hummingbird	*Leucippus baeri*	A, D	*L. taczanowskii*	dry central Peru
Short-tailed Woodstar	*Myrmia micrura*	A, D, S	?	-
Esmeraldas Woodstar (EW)	*Acestrura berlepschi*	S, E	*Acestrura* spp.	humid Andes
Ecuadorian Piculet	*Picumnus sclateri*	D, S	*P. cirrhatus*	humid E lowlands
Surf Cinclodes+	*Cinclodes taczanowskii*	A	*C. nigrofumosus*	coastal Chile
Coastal Miner+	*Geositta peruviana*	A	?	-
Necklaced Spinetail+	*Synallaxis stictothorax*	A, D	?	-
Blackish-headed Spinetail (BHS)	*Synallaxis tithys*	D, S	?	-
Rufous-necked Foliage-gleaner (RNFG)	*Syndactyla ruficollis*	S, E	*Syndactyla* spp.	humid Andes
Henna-hooded Foliage-gleaner (HHFG)	*Hylocryptus erythrocephalus*	D, S, E	*H. rectirostris*	dry SE lowlands, Brazil
Collared Antshrike+	*Sakesphorus bernadi*	A, D	*S. canadensis*	dry E lowlands
Chapman's Antshrike	*Thamnophilus zarumae*	D, S, E	*T. doliatus*	widespread, humid lowlands
Grey-headed Antbird (GHA)	*Myrmeciza griseiceps*	S, E	?	--
Watkins's Antpitta	*Grallaria watkinsi*	D, S	*G. ruficapilla*	humid Andes
Elegant Crescent-chest	*Melanopareia elegans*	A	*M. maranonica*	Marañón endemic, dry scrub
Grey-and-white Tyrannulet	*Pseudelania leucospodia*	A	?	-
Pacific Elaenia	*Myiopagis subplacens*	D, S	?	-
Pacific Royal-flycatcher (PRF)	*Onychorhynchus occidentalis*	D, S, E	*O. coronatus*	widespread, humid lowlands
Grey-breasted Flycatcher+ (GBF)	*Lathrotriccus griseipectus*	D, S, E	*L. euleri*	humid E lowlands

cont.

Table 3 (cont.)

Piura Chat-tyrant (PCT)	*Ochthoeca piurae*	A	*O. leucophrys*	dry Andes
Tumbes Tyrant (TT)	*Tumbezia salvini*	A	?	-
Ochraceous Attila+ (OA)	*Attila torridus*	S, E, W	*A. cinnamomeus?*	humid E lowlands
Rufous Flycatcher	*Myiarchus semirufus*	A	?	-
Sooty-crowned Flycatcher+	*Myiarchus phaeocephalus*	A, D	?	-
Baird's Flycatcher	*Myiodynastes bairdi*	A	*M. maculatus*	widespread, humid lowlands
Slaty Becard+ (SB)	*Pachyramphus spodiurus*	D, S	*P. rufus?*	dry E lowlands
Peruvian Plantcutter (PP)	*Phytotoma raimondii*	A	*P. rara?*	humid S. Andes
White-tailed Jay	*Cyanocorax mystacalis*	D, S	?	-
Superciliated Wren	*Thryothorus superciliaris*	A, D	*T. leucotis*	humid E lowlands
Plumbeous-backed Thrush	*Turdus revei*	D, S	?	-
Ecuadorian Thrush	*Turdus maculirostris*	D, S, E	*T. nudigens*	dry N lowlands
Three-banded Warbler	*Basileuterus trifasciatus*	D, S, E	*B. culcivorus*	humid Andes
Grey-and-gold Warbler	*Basileuterus fraseri*	D, S	*B. coronatus*	humid Andes
Black-cowled Saltator	*Saltator nigriceps*	S, E	*S. aurantiirostris*	S Andes, dry habitats
White-headed Brush-Finch	*Atlapetes albiceps*	A, D	?	(together with *A. nationi* and *A. rufigenis*,
Bay-crowned Brush-Finch	*Atlapetes seebohmi*	S, E	?	this species group is restricted to the arid
Pale-headed Brush-Finch (PHBF)	*Atlapetes pallidiceps*	D, S	?	central Andean slope)
Drab Seedeater+	*Sporophila simplex*	A	*Sporophila* spp.	widespread, dry habitats
Sulphur-throated Finch	*Sicalis taczanowskii*	A	*Sicalis* spp.	widespread, dry habitats
Cinereous Finch	*Piezorhina cinerea*	A	?	-
Crimson Finch-Tanager	*Rhodospingus cruentus*	A, D	*Coryphospingus*	widespread, lowlands
Black-capped Sparrow+	*Arremon abeillei*	A, D, S	?	-
Tumbes Sparrow	*Aimophila stolzmanni*	A, D	?	-
White-edged Oriole	*Icterus graceannae*	D, S	*I. mesomelas?*	W. Colombia-Peru
Saffron Siskin (SS)	*Carduelis siemiradzkii*	A, D	*C. magellanica*	widespread, dry habitats

The broad habitat classification distinguishes A: arid scrub and woodland, D: dry (deciduous) forest, S: semi-evergreen forest, E: evergreen forest, W: wet (cloud) forest in El Oro and Azuay Provinces in Ecuador. Phylogenetic assumptions are purely working hypotheses based on Goodwin (1970), Meyer de Schauensee (1970), Ridgely and Tudor (1989), Fjeldså (1992), P. Coopmans *in litt.* (1993), T. S. Schulenberg *in litt.* (1993) and personal knowledge. The sign "+" denotes species which are shared with another EBA. For species not listed in Ridgely and Greenfield (in prep.) taxonomy follows Meyer de Schauensee (1970). Although not considered a Tumbesian endemic by Stattersfield *et al.* (in prep.) *Attila torridus* (confined to the Tumbesian region and extreme south-western Colombia) is discussed in detail in this book as it is threatened and its range is centered on the Tumbesian region.
Abbreviations appearing after the species names refer to those used in Figures 75 and 87, and Table 13

explanations could be past allopatry, or parapatric speciation under conditions similar to today's. Parapatric speciation together with current ecological factors explain much of today's diversity (Salo 1987): for example, Gentry (1989) estimated that 75% of locally-endemic woody plant species in Amazonia represent examples of parapatric speciation.

The ideas expressed above need not rule out a connection between the dry forest areas of the Neotropics. Instead they indicate that today's species assemblage in the Tumbesian region originated from a wide variety of sources and occupied the Tumbesian region at different stages. The very distinctive species (e.g. the endemic fringillid genera *Rhodospingus* and *Piezorhina*) suggest a long evolutionary history, but disjunct populations of little-differentiated species widespread throughout arid areas in South America (e.g. *Fluvicola nengeta, Sicalis flaveola*)(Haffer 1985) are indicative of recent vicariance events. Similar conclusions were reached by Cracraft and Prum (1988) in a cladistic analysis of disjunct trans-Andean bird distributions.

CONCLUDING REMARKS

It seems likely that during the Pleistocene the Tumbesian region was more stable than many other areas. It is therefore more likely to remain similarly stable through future climatic changes (e.g. 'global warming'). The importance of this assumption for conservation actions cannot be stressed too greatly; in many areas future climatic changes may lead to the disappearance of the threatened biotic communities from reserves established specifically for their protection (Fjeldså 1991). In the Tumbesian region, climatic changes will probably lead only to a slight shift in the location of vegetation types. This will therefore allow preservation of most flora and fauna within areas designated as important today. Large reserves (especially the North-West Peru Biosphere Reserve) stand out in this respect, as they contain a wide range of vegetation types which would permit some shifts should climatic changes come.

■

VEGETATION

INTRODUCTION

IN ECUADOR and northern Peru, where botanical exploration has been concentrated in the highlands and the Amazonian rainforests, the dry and humid forests of the Tumbesian region are among the least known regions botanically.

While this chapter will focus on the entire Tumbesian region, most descriptions and analyses will concentrate on the area visited by M. Kessler during recent fieldwork (Kessler 1992): the western Andean slopes below 2,000 m in the provinces of Loja and El Oro in south-western Ecuador, including adjacent areas of Azuay Province (north to 3°S). This region will from now on be referred to as "south-west Ecuador". The validity of extrapolating conclusions obtained from the study of this rather small section of the Tumbesian region to the whole centre of endemism will be discussed. Conservation recommendations will also focus on the Tumbesian region as a whole.

PREVIOUS STUDIES

So far very little has been published on the phytogeography of Ecuador and particularly its south-western part. Even though the science of phytogeography itself originated in Ecuador when Alexander von Humboldt and A. Bonpland completed their *Ideen zu einer geographie der pflanzen* [Ideas on the Geography of Plants](published in 1807), surprisingly little detailed attention has been paid to the vegetation patterns in the present century. Chapman (1926) established four Life Zones (actually altitudinal zones) based on his ornithological work; Diels (1937) subdivided Ecuador into 10 phytogeographic formations, and in *The forests of western and central Ecuador* Holdridge and Little (1944) described eight forest types. However, of those classifications none describes more than five vegetation types for south-west Ecuador and they are therefore of little use for any detailed analysis of the distribution and ecology of individual plant and animal species.

In 1968 Acosta-Solis proposed a more detailed geobotanical classification of

Ecuador but again the scale is too small for such an area as south-west Ecuador. In 1978 the Ecuadorian Ministry of Agriculture and Farming produced 1:1,000,000 bioclimatic and ecological maps in collaboration with the French Office of Scientific and Technical Research (Cañadas Cruz and Estrada 1978, Cañadas Cruz 1983). Since then a series of 1:200,000 ecological maps has been published by these institutions. These are based on the rather rigid classification system of Holdridge (1956) discussed on page 100.

Despite various collecting expeditions to the Ecuadorian provinces of El Oro and Loja, few researchers have used a phytogeographical basis to describe the flora. Svenson (1946) and Valverde *et al.* (1979) described the flora of the arid Santa Elena Peninsula in Guayas Province, giving some consideration to bioclimatic factors, and in the 1980s Danish botanists (mainly J. Madsen) started detailed studies in the dry forests on Puná Island in the Gulf of Guayaquil. The botanical information from three ethnobotanical expeditions to Loja Province (Ecuador) and Piura Department (Peru) between 1959 and 1989 was published in 1990 (Empraire and Friedberg 1990). Munday and Munday (1991) described the vegetation of a limited area around Sozoranga, Utuana and Macará in Loja Province, Ecuador. A rather complete treatment of the vegetation of the Chongón-Colonche Hills (Guayas Province) was presented by Valverde (1991). Further botanical exploration was conducted in 1991 at various sites in western Ecuador by Conservation International's Rapid Assessment Programme (Parker and Carr 1992) and is being continued by C. Josse. At the time of writing Fundación Natura is conducting a detailed study of the Very Humid Lower Montane Cloud-forests around Molleturo in Azuay Province, Ecuador. The deciduous tropical forest will be treated in Dodson and Gentry's *Flora of Capeira* (in prep.) a locality near Guayaquil in the province of Guayas. A study of the vegetation and human impact in Loja Province was conducted by Wann (1992).

The situation is somewhat better for Peru. Most important is Weberbauer's (1911, 1945) monumental treatment of the Peruvian flora, still the standard reference. Subsequently the western Andean slopes were studied in detail by H. W. and M. Koepcke, leading to the publication of a general synecological and biogeographical treatment of the region (Koepcke 1961). A life zone map (1:1,000,000) based on the Holdridge system (1956) was published by Tosi in 1960. Subsequently most research in the departments of Tumbes and Piura was conducted by foresters; only recently have C. Días and A. H. Gentry, and CDC-UNALM (1992) studied the dry forests of the Cerros de Amotape National Park on a general botanical basis.

THE VEGETATION TYPES OF THE TUMBESIAN REGION

Material

In order to prepare the vegetation classification which follows, fieldwork was carried out in south-west Ecuador in early 1991, details of which can be found in Kessler (1992). This central part of the region was chosen because it was not possible to study the entire Tumbesian region in depth. Therefore, the vegetation classification concentrates on this area where the diversity of different vegetation types is highest. Other data used in this chapter include published and unpublished vegetation descriptions and data from herbarium specimens (mainly from the herbarium of the Catholic University, Quito and the Herbarium Jutlandicum, Aarhus, Denmark). Climatic data were obtained from Munday and Munday (1992).

Method of vegetation classification

The classification of vegetation types has always depended on the personal judgement of particular botanists, some relying mainly on floristic comparisons (e.g. Braun Blanquet 1921), others on climatic data (e.g. Holdridge 1956), and others still on a combination of both, including the 'physiognomic aspect' (Mueller-Dombois and Ellenberg 1974). The approach used in the present classification most closely resembles that of Mueller-Dombois and Ellenberg (1974). Rather than claiming objectivity in a field which has so far defied clear definitions for vegetation types, it should be recognized that the proposed classification is based on the authors' subjective impression of what constitutes a separate vegetation type. This impression was based primarily on physiognomy (i.e. overall appearance), but floristic composition and climate also contributed. The classification is therefore pragmatic, intended for use to interpret the distribution of particular plant and animal species.

Another difficulty concerns human impact. Generally we have tried to (re)construct the potential natural vegetation, which will be described below. Often patches of remnant vegetation are found which allow interpretation of the observed vegetation patterns, but in some areas (especially the arid ones) only tiny fragments (e.g. in steep ravines) can be used to estimate what the original vegetation might have looked like. However, advancing habitat destruction often demanded an arbitrary decision to be taken now; it is recognized that others might have chosen differently.

Often limits dividing different vegetation types were observed while travelling through the study area by vehicle (Figure 14), making conclusions somewhat tentative. Where two vegetation types appeared to merge, simplicity demanded that a choice be made to avoid confusion due to intermediate gradations between quite distinct types. It should thus be borne in mind that the description of a vegetation type focuses on its 'typical' aspect which can be strongly modified as another vegetation type with differing ecological characteristics is approached.

Figure 14. Study sites and travel routes (heavy lines) in south-west Ecuador used to compile the current classification. Main study sites: A: Sozoranga, Q. Suquinda; B: Tambo Negro, C: Tierra Colorada, D: Hacienda Buenaventura. Sites visited briefly: 1: Celica; 2: Río Catamayo valley near El Limon; 3: Catacocha; 4: Río Amarillo valley 2 km S of Portovelo; 5: El Portete E of Piñas; 6: Uzhcurrumi; 7: Río Jubones valley at 1,000 m; 8-9: Río Rircay valley (1,500-1,800 m); 10: Río Leones valley below Oña.

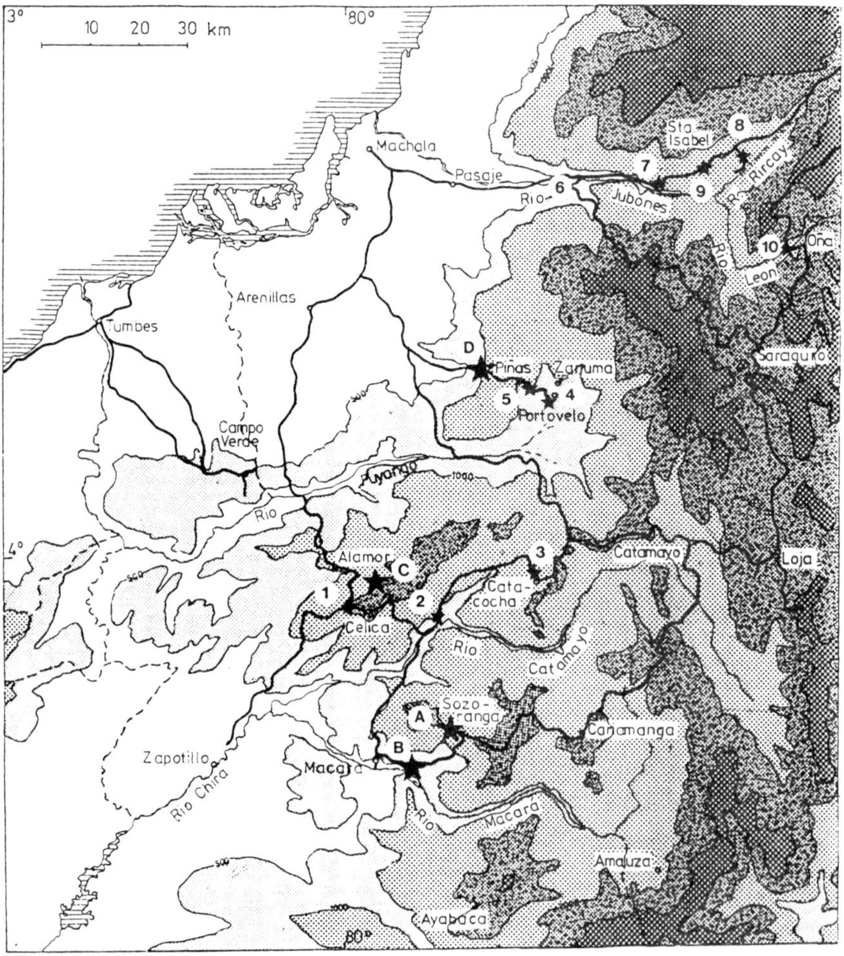

Box 5. Vegetation terms used in this book.

Forest Woody vegetation (usually trees) at least 2 m tall and with at least 25% coverage. Forests taller than 20 m are called *high* forest, between 10 and 20 m *moderately high* forest, between 5 and 10 m *low* forest and below 5 m *dwarf* forest.

Scrub Woody vegetation 0.2-2 m tall.

Forest and scrub can be *closed* (81-100% coverage, crowns usually interlocking), *open* (41-80% coverage) or *very open* (25-40% coverage). Those dominated by spiny or thorny species are called *thorny*.

Deciduousness
Mainly deciduous: no completely deciduous vegetation was found during fieldwork in south-west Ecuador. Instead a few trees (e.g. *Ceiba* or *Acacia*) keep their leaves through much of the dry season.

Semi-evergreen: in this case a considerable proportion of the trees shed their leaves in the dry season, leaving enough trees and shrubs with leaves to maintain the green appearance of the forest.

Moisture The terms humid and very humid are used to distinguish between different evergreen forest types.

Humid: with a mean annual precipitation in the range 1,500-2,500 mm.
Very humid: with higher precipitation (to 3,000+ mm/year) and often receiving additional moisture from frequent mist.

Cloud We follow Stadtmüller (1987) who defined cloud-forests as "all forests in the humid tropics that are frequently covered in clouds or mist; thus receiving additional humidity, other than rainfall, through the capture and/or condensation of water droplets (horizontal precipitation), which influences the hydrological regime, radiation balance, and several other climatic, edaphic and ecological parameters".

Altitude In addition to stating the altitude above sea-level of a particular site, this may also give information on the temperature regime of the site.

Lowland (tropical): upper limit varies between 400 and 600 m; mean annual temperatures in south-west Ecuador are 23-26°C.
Premontane: lower limit is 400-600 m, upper limit ranges from 1,000 to 1,200 m, mean annual temperature is 20.5-23°C, up to 24°C at intermontane localities.
Lower Montane: altitudinal range 1,000-1,200 m to 1,500-1,800 m; mean annual temperatures 18(19)-20.5(21)°C.
Montane: above 1,500 to 1,800 m; mean annual temperature below 18°C (to 16°C at 2,000 m near Celica in Loja Province).

Floristic components In some cases characteristic or dominant species are included in the name.

Nomenclature

Names given mostly follow an unpublished system designed by H. Ellenberg, S. and G. Miehe, C. Leuschner, I. Hensen and M. Kessler in 1992. For the sake of simplicity some names do not include the full array of possible terms. For example, it is clear that all deciduous forests in the study area are drought-deciduous and thus it seems reasonable just to call them deciduous. Terms for naming the vegetation types are defined in Box 5.

The vegetation types of Loja, El Oro and adjacent Azuay Provinces, south-west Ecuador

The habitat types are arranged by increasing humidity of habitat type within each altitudinal zone. Often information from outside the area is included; in some cases notes are made on species composition or the appearance of particular vegetation types in regions outside south-west Ecuador. The climatic data and the limits of altitude and precipitation presented with each vegetation type only apply to south-west Ecuador and can be quite different in other parts of the Tumbesian region.

MAINLY DECIDUOUS TROPICAL THORN-FOREST AND *ACACIA* THORN-FOREST

(Natural) limits	lower	upper
elevation (m)	0-50	50-400
mean annual precipitation (mm)	100-200	500

Climate Characterized by low precipitation, high temperatures (mean for Zapotillo 26.2°C), a long dry season (May-December at Zapotillo) and little additional moisture from fog condensation.

This formation poses a particular problem. Although *Acacia* thorn-forest probably occurs naturally in south-west Ecuador, it also is *the* common substitute vegetation in (mainly heavily degraded) arid areas, resulting in large areas of the intermontane valleys being *Acacia*-dominated. It can replace both *Ceiba* forest types, arid intermontane forest and semi-evergreen forests, but on the other hand natural thorn-forest is readily degraded to a desert-like vegetation with very few shrubs. The natural species composition and distribution of the *Acacia* Thorn-forest can often be cause for conjecture; here it is considered to have naturally occurred on the lower slopes towards Peru (often dominated there by *Prosopis*) and probably in the driest areas of the Jubones and Catamayo valleys (see under arid intermontane forest).

Mainly Deciduous Tropical Thorn-forest is usually only about 5 to 10 m high, rather open and with a deciduous herb layer. In the dry season therefore it is characterized by a few evergreen and a majority of dormant trees and shrubs; in the rainy season a dense, high herb layer conveys an aspect of lushness (Figure 15). This is the only vegetation type in south-west Ecuador (with the exception of the high montane Páramos) where grasses are prominent among the herbs.

Figure 15. Mainly Deciduous Tropical Thorn-forest and *Acacia* Thorn-forest. Forest dominated by *Loxopterygium* and *Acacia* along a dry stream bed in the rainy season. Note the abundance of low shrubs (*Ipomoea*) and vines. Between El Ceibo and Sabanilla, Loja Province, Ecuador (350 m, 18 February 1991).

Species composition
Species found on a very brief visit to El Ceibo on the Celica-Zapotillo road included:

Acacia macracantha
Prosopis juliflora
Loxopterygium huasango
Capparis angulata
Pithecellobium excelsum
Armatocereus cartwrightianus
Ipomoea carnea

More detailed descriptions by Weberbauer (1945:280-284) from adjacent Piura and Tumbes Departments in Peru also include (species marked with an asterisk are found in areas with groundwater influence):

Trees:

Capparis mollis
Caesalpinia corymbosa
Bursera gravolens
Celtis iguanea *
Zizyphus piurenses *
Muntingia calabura

Shrubs:

Mimosa acantholoba
Cercidium praecox
Cordia rotundifolia
Cryptocarpus pyriformis
Capparis cordata
C. ovalifolia
Vallesia dichotoma
Monnina pterocarpa
Waltheria sp.
Jaquemontia sp.
Ruellia sp.
Althernanthera sp.
Isocarpha microcephala
Maytenus orbicularis
Acacia tortuosa
Parkinsonia aculeata
Grabowskia boerhaviifolia
Galvesia limensis
Encelia canescens
Cereus macrostilbas

Herbs and Grasses:

Amaranthus sp.
Heliotropium angiospermum
Nicandra physaloides
Datura sp.
Physalis 2 spp.
Solanum several spp.
Sicyos chaetocephalus
Schizoptera trichotoma
Luffa operculata
Eragrostis sp.

Bouteloua disticha
Aristida adscensionensis
Chloris virgata
Anthephora hermaphrodita

Near its southern limit in the department of Lambayeque in Peru this forest type is restricted to altitudes between 400 and 900 m on the Andean foothills; its species composition changes somewhat and is dominated by *Celtis iguanacea, Pithecellobium multiflorum, Acacia macracantha, Loxopterygium huasango, Caesalpinia papai, Bursera graveolens* and *Ficus* spp. (de Macedo 1979).

MAINLY DECIDUOUS *CEIBA TRICHISTANDRA* FOREST

Natural limits	**lower**	**upper**
elevation (m)	0-400	150-1,400
mean annual precipitation (mm)	200-500	500-1,100

Climate Coastal areas have rather little precipitation (200-500 mm) and a seven-month-long potential dry season, but its effect is ameliorated by frequent fog and cloud cover at this time. Forests without this additional moisture need higher precipitation (350-1,100 mm). Mean annual temperatures range from 20 to 25°C.

Physiognomically the most unmistakeable forest type in south-west Ecuador, the *Ceiba trichistandra* forests are marked by the predominance of bottle-bellied bombacaeous trees of *C. trichistandra* and *Eriotheca ruizii* and the spiny, red-barked Fabaceae *Erythrina velutina*. With a large ecological amplitude and an altitudinal range of 1,500 m, it shows a somewhat different species composition at different sites (e.g. the Puyango valley as compared to the Catamayo valley); a more detailed study might subdivide this forest type into coastal (tropical) and intermontane (premontane) types. The *Ceiba* forest was called “savanna” by Harling (1979), but it is certainly a forest type.

Undisturbed *Ceiba* forests are 20 to 25 m tall with single trees reaching 35 m. Tree crowns touch and species are mostly drought-deciduous (except the *Ceiba* trees themselves)(Figure 16). There are well developed layers of subcanopy trees and shrubs of which some of the constituent species are partially evergreen and have their flowering or fruiting seasons in the dry season (e.g. *Ceiba, Cochlospermum, Ficus, Armatocereus*), a matter of vital importance for the survival of frugivorous birds and mammals. The herb layer is completely deciduous and very prominent in the rainy season, though grasses are uncommon.

Figure 16. Disturbed Mainly Deciduous *Ceiba trichistandra* Forest.
Two views of the same place (Tambo Negro, Loja Province, Ecuador) on 3 February 1991 (left), two days after the onset of the annual rains, and on 8 March 1991 (right), five weeks later. The tree in the centre is *Cochlospermum vitifolium*, those with thick trunks in the background are *Ceiba trichistandra*.

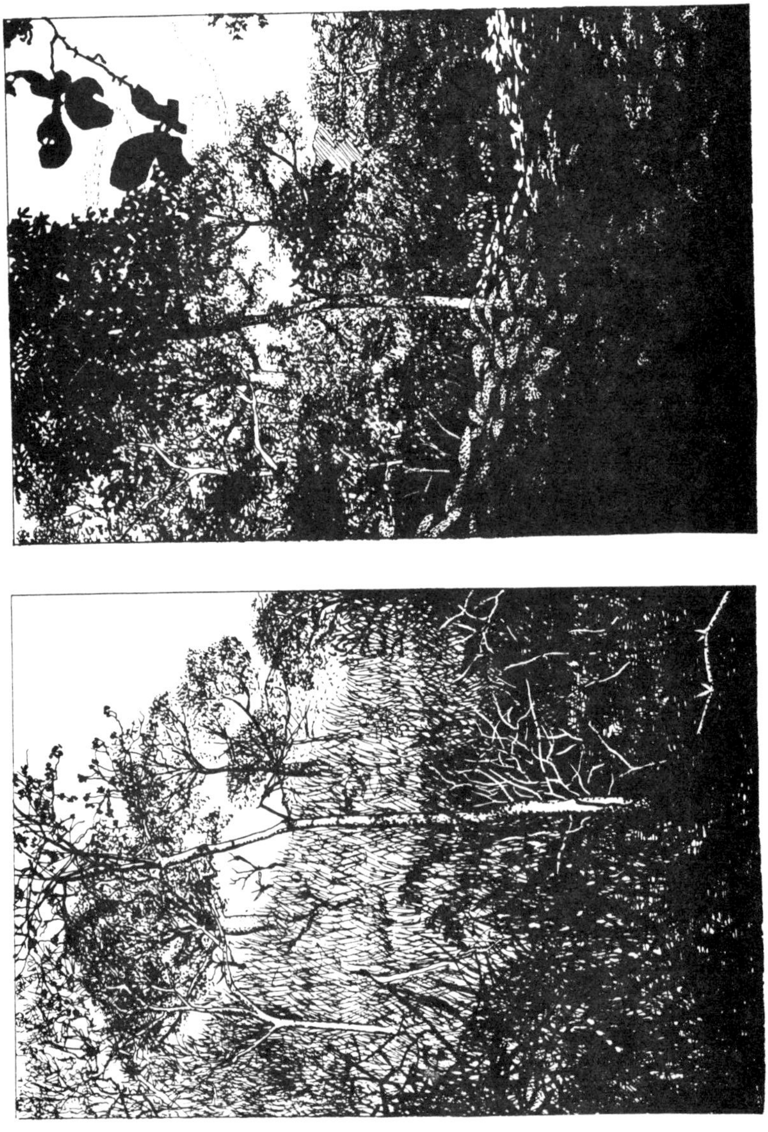

Canopy cover was estimated at Tambo Negro in Loja Province on 3 February 1991 (two days after the start of the annual rains) and on 8 March 1991, from a series of ten photographs taken with a 28-mm wide-angle lens from points 20 m apart along a trail that runs about 50 m above the river and parallel to it. Canopy cover in the dry season was 17% plus or minus 11% (Figure 17, top); eight weeks later it had risen to 50% plus or minus 13% (Figure 17, bottom). Riverine forests are described as azonal vegetation types on page 92.

Figure 17. Mainly Deciduous *Ceiba trichistandra* Forest.
Top: view into canopy at Tambo Negro, Loja Province, Ecuador, on 3 February 1991. Bottom: same view on 8 March 1991. Tree in lower right corner is *Ceiba trichistandra*; that with long branches and short leaves *Pithecellobium exelsum*. Note that a few bromeliads are visible at the end of the dry season.

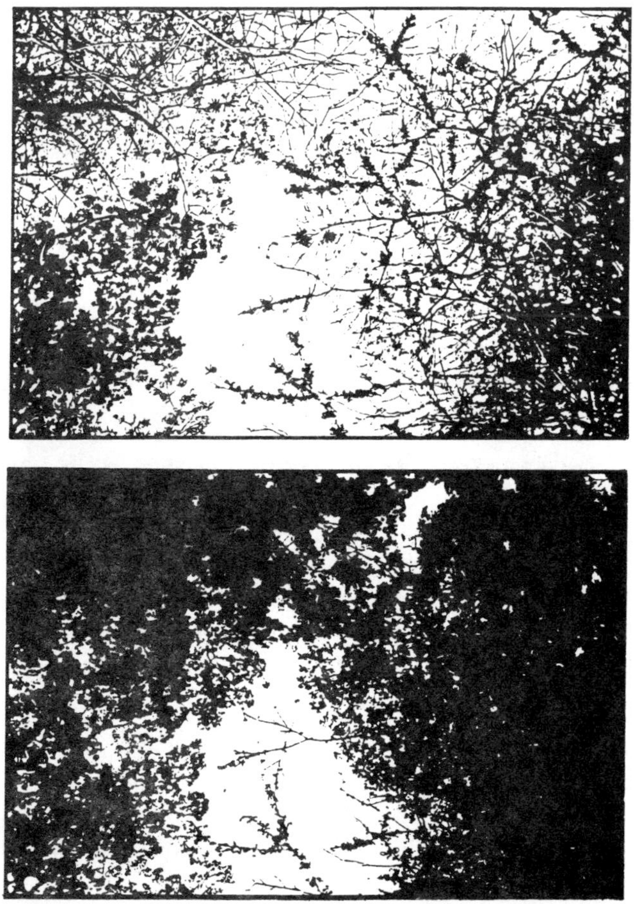

Species found at Tambo Negro include:

Canopy trees:

Ceiba trichistandra
Eriotheca ruizii
Erythrina velutina
Cochlospermum vitifolium
Tabebuia chrysantha
Hura crepitans (near river)
div. Leguminosae

In Peru *Eriotheca ruizii* is replaced by *E. discolor* (Weberbauer 1945).

Understorey trees and shrubs:

Heliotropium cf. *angiospermum*
Capparis hetrophylla
Urera baccifera
Senna atomaria
Achatocarpus nigricans
Cornutia sp.
Randia sp.
Pithecellobium excelsum
Ficus americana
Triplaris cumingiana
Cordia lutea
Caesalpinia sp.
Zizyphus cf. *thyrsiflora*
Guazuma ulmifolia
Albizia multiflora
Croton fraseri
Annona sp.
Trichilia hirta
Senna mollissima
Phyllanthus anisolobus
Rauvolfia tetraphylla
Acnistus cf. *arborescens*
Carica parviflora
Armatocereus cartwrightianus

Disturbed and heavily grazed areas are dominated by spiny shrubs and vines (Figure 18):

Acacia macracantha
Bougainvillea peruviana
Pithecellobium excelsum
Prestonia mollis
Mansoa hymenaea
Macranthisiphon longiflorum
Arrabidaea corralina

Figure 18. Dense *Acacia macracantha* thicket in heavily disturbed Mainly Deciduous *Ceiba trichistandra* Forest. Tambo Negro, Loja Province, Ecuador (600 m, 29 January 1991).

During the rainy season the ground is covered by annual herbs and geophytes:

cf. *Acalypha setosa*
Coursetia caribaea
Salvia cf. *perlucida*
Adiantum raddianum
Talinum cf. *triangulare*
T. paniculatum
Commelina cf. *erecta*
Stenomesson sp.
Mirabilis cf. *violacea*
Brachiaria fasciculata

Harling (1979) further lists the cucurbit climbers *Apodanthera, Sicyos, Momordica* and *Luffa*, and the grasses *Pennisetum purpureum, P. occidentale, Aristida adscensionensis, Panicum* spp., *Paspalum* spp., *Andropogon bicornis* and *Chloris virgata*.

Epiphytes are quite common but not very diverse and dominated by succulent bromeliads, cacti and orchids. *Erythrina velutina* seems to present the most favourable conditions for bromeliads. On exposed ridges *Tillandsia usneoides* is particularly common.

Bromeliads:

Guzmania monostachya
Tillandsia caerulea
T. capillaris
T. complanata
T. disticha
T. latifolia var. *divaricata*
T. flagellata
T. floribunda
T. multiflora
T. triglochinoides
T. usneoides
Pitcairnia prolifera
Vriesea espinosae

Near Macará Rauh (1984) further recorded *Tillandsia lehmannii, T. marnieri-lapostollei* and *Vriesea barclayana* and lists these and the above species as typical of the *Ceiba* forests from Guayas Province, Ecuador to northern Peru.

Other epiphytes include:

Rhipsalis micrantha
Hylocereus cf. *polyrhizus*
Epidendrum sp.
Trichocentrum tigrinum
Polypodium cf. *polypodioides*
Peperomia sp.
Philodendron sp.

SEMI-EVERGREEN *CEIBA PENTANDRA* FOREST

(Natural) limits	lower	upper
elevation (m)	0-1,000	100-1,200
mean annual precipitation (mm)	500-1,000	1,000-1,300

Climate In addition to the vertical precipitation this forest type generally receives additional moisture from horizontal precipitation that lessens the effect of the six-month-long potential dry season. Temperatures lie in the range 23 to 26°C.

Especially in the dry season this forest type is readily distinguished from its drier, more seasonal counterpart. Rather than being dominated by a few characteristic species this forest is composed of a large number of superficially similar species with only *Ceiba pentandra*, the widespread Kapok tree, standing out (Figure 18). A considerable portion of the canopy trees are evergreen, so distinguishing this forest type apart from *Ceiba trichistandra* forest. On the other hand palms, Cyclanthaceae, Heliconias or Melastomataceae, all characteristic of humid, evergreen forests, are absent.

The rare extant examples of undisturbed *Ceiba pentandra* forest are about 20 to 30 m tall with large *Ceiba* trees reaching 50 m or more. The structure is similar to that of the *C. trichistandra* forest with numerous understorey trees and shrubs, most of which are evergreen. Herbs are rare; epiphytes fairly common.

Figure 19. Semi-evergreen *Ceiba pentandra* Forest.
Relict patch on steep slope. The *Ceiba* is the tall tree with light-coloured trunk. Between río Puyango and Palmar, El Oro Province, Ecuador (650 m, 20 February 1991).

Species
Tosi (1960) lists the following tree genera for the comparable "subtropical dry forest" of the Tumbes National Forest, Peru:

Alseis
Citharexylum
Cochlospermum
Caesalpinia
Centrolobium
Colubrina
Dalea
Embothrium
Erythrina
Fagara
Inga
Jacaranda
Ladenbergia
Linoceria
Phyllanthus
Pithecellobium
Roupala
Tecoma
Tabebuia
Schinus
Sapindus

SEMI-EVERGREEN LOWLAND AND PREMONTANE TALL FOREST

(Natural) limits	lower	upper
elevation (m)	0-1,000	400-1,400
mean annual precipitation (mm)	900-1,300	1,400-1,700

Climate Similar to that of the preceding vegetation type, but with more precipitation, lower temperatures (estimated at 19 to 24°C) and a shorter dry season (three to six months).

This is the somewhat more humid version of the previous forest type. Most species are evergreen with drought-adapted species such as *Ceiba* lacking. The favourable climate and a rather flat topography have made the lower areas of this vegetation type victims of monoculture (extensive banana and cocoa plantations), leaving only tiny forest fragments, none of which have been visited by the authors. Larger areas remain on the Andean foothills, mainly in Peru, but these too have been only studied superficially.

The Semi-evergreen Lowland and Premontane Tall Forests are about 20 to 30m high; occasionally 40 m tall trees can be found. As only a fraction of the trees are drought-deciduous the forest keeps its green appearance throughout the year and is able to support large folivorous and frugivorous mammals (the monkeys *Alouatta* and *Cebus*). Buttressing and strangler figs are not uncommon.

MOIST LOWLAND FOREST

(Natural) limits	lower	upper
elevation (m)	150-300	500-600
mean annual precipitation (mm)	1,100	2,300

This evergreen forest type borders the Tumbesian dry forests in the north and extends along the Andean foothills to the north of El Oro Province. Since it does not belong to the Tumbesian region, and as it has been studied in detail by Dodson *et al.* (1985) at Jauneche (Los Ríos Province), it will not be treated in detail here. It is a rather high (30 m) forest type which contains numerous deciduous tree species but which is basically evergreen. Dodson *et al.* (1985) list 728 vascular plants species for the 134 ha forest patch near Jauneche.

HUMID TO VERY HUMID PREMONTANE CLOUD-FOREST

(Natural) limits	lower	upper
elevation (m)	500-600	1,100-1,500
mean annual precipitation (mm)	1,400-1,700	3,000+

Climate Even though the precipitation data might suggest a dry period of a few months, this vegetation type has a constantly humid climate due to the frequent condensation of clouds on the seaward mountain slopes to which this forest type is restricted. Mean annual temperature is in the range 19 to 23°C.

The special topographical and climatic conditions which produce this localized type of very humid forest will be discussed below. Abundant epiphyte growth, often forming dense mats on the branches of canopy trees, is a characteristic and striking feature (Figure 20). Epiphyllous fungi, lichens, mosses and liverworts are also common. In the Piñas area this vegetation type occurs above the semi-evergreen lowland forest and is readily distinguished by the occurrence of numerous Melastomataceae, Zingiberaceae, Cyatheaceae (tree-ferns) and such characteristic palms as *Iriartea, Euterpe* and *Phytelephas*.

This forest type is similar to the "Wolkenwald" described by Vareschi (1980: 104-127) who lists only two examples, the Cameron highlands on the Malakka Peninsula in South-East Asia and the northern coastal cordillera in Venezuela (Rancho Grande), but predicts that a few additional areas will be found.

Trees are tall, averaging 30 or 35 m, occasionally reaching 45 m. Larger trees often lack branches below 20 m and have rather flat crowns; slight buttressing is common. In the forests studied at Buenaventura, tree crowns and the canopy were rather open, thus allowing for a dense middle- and understorey. It is questionable whether this is a natural state or the result of selective logging.

The following species list has been compiled from personal observations and

Figure 20. Very Humid Premontane Cloud-forest.
Note the rather open canopy, dense undergrowth and dense mats of epiphytes on tall tree in background. Hacienda Buenaventura, El Oro Province, Ecuador (900 m, 26 February 1991).

collections at Buenaventura and from a preliminary list by Fundación Natura (April 1991) of trees from the Molleturo area in Azuay Province. Species listed only from Molleturo are not marked; those marked with one asterisk were found at both localities and those with two asterisks only at Buenaventura. It should be remembered that the Molleturo list includes species from a wide altitudinal range (100-2,000 m).

Actinidiaceae
*Saurauia** cf. *bullosa*
Anacardiaceae
Spondias purpurea
Annonaceae
Annona cf. *edulis*
*Guatteria**
Rollinia
Aquifoliaceae *Ilex*
Araliaceae
Oreopanax
Dendropanax
Arecaceae (Palmae)
Euterpe cf. *precatoria**
*Iriartea deltoidea**
Prestoa cf. *asplundii*
*Geonoma**
Chamaedorea cf. *poeppigiana*
*C. pinnatifrons***
*Phytelephas aequatorialis**
Asteraceae
Baccharis
Vernonia baccharoides
Berberidaceae *Berberis*
Bignoniaceae
Delostoma
Tabebuia guayacan
Bombacaceae
*Matisia cordata**
*Ochroma pyramidale**
Pachira
Ceiba pentandra
Boraginaceae *Cordia alliodora*
Brunelliaceae *Brunellia*
Burseraceae
Tetragastris
Dacryodes peruviana
Caesalpiniaceae *Bauhinia**
Caprifoliaceae *Viburnum*
Caricaceae
*Jacaratia spinosa**
*Podandrogyne brevipedunculata***
Cecropiaceae *Cecropia**
Celastraceae *Maytenus*
Chlorantaceae
Licanea
*Hedyosmum**
Clusiaceae
*Clusia** cf. *dixonii*
C. cf. *multiflora*
C. cf. *laxiflora*
C. cf. *alata*
Symphonia globulifera
Tovomitopsis
*Rheedia edulis**
Chrysochlamis
Cunoniaceae
*Weinmannia**
Elaeocarpaceae
Vallea
Muntingia calabura
Ericaceae *Psammisia***
Euphorbiaceae
Hyeronima
Alchornea
Mabea occidentalis
*Sapium**
*Cordia cylindrostachya***
Acalypha cf. *macrostachya***
Fabaceae
*Erythrina**
Centrolobium paraense
Flacourtiaceae
*Casearia** *mariquitensis*

Banara guianensis
Gesneriacea
*Alloplectus teuscheri***
*Besleria***
*Columnea***
Hippocrateaceae *Salacia*
Icacinaceae *Calatola*
Lauraceae*
Endlicheria
Nectandra
Lecythidaceae
*Eschweilera**
*Gustavia**
Melastomataceae
*Miconia rivalis***
*M. brevitheca***
*M. centrodesmoides***
*M. dodecandra***
*M. goniostigma***
*M. fosbergii***
M. sp. nov.?
(aff. *andersonii, fosbergii)***
M. macrotis var. *canescens*** (600 m)
Miconia spp.*
Meriana
Topobaea cf. *pittieri***
*Conostegia cuatrecasii***
*C. centronoioides***
*Blakea subconnata***
*Ossaea boeckii***
*O. micrantha*** (600 m)
Leandra cf. *dichotoma*** (600 m)
*Graffenrieda cucullata***
*Arthrostemma ciliatum***
*Clidemia hirta***
*Triolena hirsuta***
*Tibouchina laxa***
*T. longifolia***
*Monochaetum lineatum***
Meliaceae
*Guarea**
Cedrela odorata*
C. montana
Axinea
*Trichilia**
Menispermaceae *Abuta*
Mimosaceae
Inga edulis*
I. spectabilis
Samanea saman
Moniniaceae *Siparuna**
Moraceae
Batacarpus
Ficus gigantorice*
F. macbridei
Brosimum latescens
Castilla elastica
Trophis racemosa
Pseudolmedia eggersii
Artocarpus altilis
Myricaceae
Myrica pubescens
Myristicaceae
Virola sebifera*
Myrsinaceae
*Myrsine**
*Ardisia**
*Cybianthus***
Myrtaceae
*Myrcianthes**
*Psidium**
Nyctaginaceae *Neea*
Papaveraceae *Bocconia integrifolia*
Piperaceae *Piper* spp.*
Polygonaceae
Triplaris
Coccoloba obovata
Proteaceae
Oreocalyx grandiflora
Lomatia hirsuta
Rosaceae
Prunus
Rubiaceae
Cinchona
Elaeagia
Palicourea
Pentagonia
Faramea

Randia cf. *armata*
Rutaceae *Zanthoxylum**
Sabiaceae *Meliosma*
Sapindaceae *Allophyllus**
Sapotaceae
*Chrysophyllum** *venezuelensis*
Pouteria
Solanaceae
Cestrum cf. *megalophyllum*
*Solanum**
Iochroma
Staphyleaceae
Huertea glandulosa
*Turpinia occidentalis***

Tiliaceae
Heliocarpus americanus
Apeiba
Trichospermum
Ulmaceae
Ampelocera
Trema
Urticaceae
*Urera caracasana***
Verbenaceae
Citharexylum
Aegiphila cf. *alba***
Violaceae *Gleospermum*

Further understorey species (some as epiphytes) recorded at Buenaventura include:

Columnea minor
Pilea aff. *marginata*
Cyathea (3 species)
Cnemidaria
Thelypteris grandis
Polypodium fraxineum
Polypodium spp.
Blechnum cordatum
Microgramma piloselloides
Dicranopteris pectinata
Hymeniphyllum sp.
Asplenium auriculatum
Athyrium dombeyi
Selaginella spp.
Polypodium fraxinifolium
Elaphoglossum sp.
Nephrolepis rivularis

Epiphytes are abundant, including lichens, mosses, liverworts, ferns, *Peperomia* spp., Araceae and numerous orchids. The bromeliads, which were studied in more detail, include:

Guzmania angustifolia
G. garciaensis
G. hitchcockiana
G. lingulata
G. monostachya
G. patula
G. wittmackii
Tillandsia acosta-solisii
T. narthecioides
T. pseudotetrantha
T. venusta
Vriesea sp.

Note the large number of *Guzmania* species as compared to the predominance of *Tillandsia* species in more arid vegetation types.

Disturbed areas along roadsides have dense thickets of:

Heliconia (3 spp.)
Costus (2 spp.)
Gleicheniaceae
Chusquea sp.
Piper spp.
Tibouchina spp.
Cecropia sp.
Sobralia spp.
Epidendrum sp.

A similar forest was studied in detail by Parker and Carr (1992) at Manta Real in Azuay Province.

DECIDUOUS TO SEMI-EVERGREEN INTERMONTANE SCRUB, THORN-FOREST AND FOREST

(Natural) limits	**lower**	**upper**
elevation (m)	500-1,000	2,000+
mean annual precipitation (mm)	150	400-800

Climate Low precipitation levels, high temperatures (22 to 26°C), long, pronounced dry seasons (e.g. eight months for La Toma in Loja Province) and almost no additional moisture from fog condensation characterize this vegetation type, which includes the most arid portion of the study area.

As so often happens in arid areas, this formation gives special difficulty, not so much because of fragmentation as in the case of humid forests, but rather because the forests are generally and gradually degraded (mostly by grazing and logging) often leaving no natural patches to aid the interpretation of the potential natural vegetation. The resulting vegetation is often scrub composed of *Acacia*, *Croton* and Dodonea. Certainly the various river valleys in south-west Ecuador support a distinctive vegetation type, but it seems questionable whether this can be reconstructed given the present state of degradation. Apparently human settlement in pre-Columbian times began in the arid valleys rather than in the humid forests, since they offer good conditions for agriculture due to the few pests and herbs they have (J. Madsen verbally 1991). As the arid regions became degraded people moved into more humid areas, a pattern that can still be observed today.

In the Jubones valley three distinct arid areas can be recognized when travelling up-river from Machala to Cuenca. After leaving the moist lowland forest around Uzhcurrumi a semi-evergreen, 5-m-high thorn-forest is found between 500 and 1,000 m+. The large tree cactus *Pilosocereus tweedyanus* was particularly conspicuous. Further up-river, in the most arid part of the Jubones valley (estimated mean annual precipitation as little as 150 mm) between 1,000 and 1,500 m, very little vegetation is found even in the rainy season. Some areas were desertic with only a few small *Acacia* trees growing in dry ravines while in others cacti and succulent species dominated (Figure 21). Species include:

Figure 21. Intermontane Desert Scrub.
Note the mostly bare ground, even in the rainy season, due to overgrazing. Columnar cacti are *Espostoa lanata,* shrubs *Euphorbia weberbaueri* and *Croton collinus*. Río Jubones valley, Azuay Province, Ecuador (1,000 m, 22 February 1991).

Espostoa lanata	*Tillandsia caerulea*
Opuntia cf. *quitensis*	*T. disticha*
O. pubescens	*T. latifolia*
Euphorbia weberbaueri	*T.* cf. *macrodactylon*
Carica sp.	*T. secunda*
Croton collinus	*T. tectorum*
Dodonaea viscosa	*Puya* sp.

Presumably these species were originally restricted to extremely xeric sites on steep sunny slopes, while the general vegetation could have been some kind of semi-evergreen (*Acacia*?) forest. No human habitation is found away from the river oasis. Above 1,500 m in the río Rircay valley there is a noticeable increase in humidity, and human occupation became suddenly apparent. Only fragments of low, degraded forest are found. A collective list from three localities between 1,500 and 1,800 m includes:

Caesalpinia spinosa	*Capparis flexuosa*
Senna bicapsularis	*Coccoloba ruiziana*
Acacia macracantha	*Cordia polyantha*
Sapindus saponaria	*Ipomoea carnea* (on degraded slopes)
Piper sp.	

and the bromeliads:

Tillandsia disticha	*T. recurvata*
T. latifolia	*T. tripinnata*

Rauh (1977) further lists *Pitcairnia heterophylla, Tillandsia mima, Vriesea olmosana* and *V. rauhii* from a locality at 1650 m.

Forest fragments in the río Leones valley at 1,800 m below Oña were up to 5 m tall (Figure 22) and contain:

Acacia macracantha	*Alnus acuminata*
Caesalpinia spinosa	*Tecoma castanifolia*
Cantua quercifolia	*Ilex* sp.

In the vicinity of the town of Catamayo the valley of the same name show a similarly dry aspect as the Jubones valley. The presence of two species of globular cacti indicate that in some areas only semi-desert could naturally grow (J. Madsen verbally 1991). At 1,300 m arid slopes are covered by heavily grazed *Acacia* and *Dodonea* scrub. Only occasional trees are found, mainly *Capparis angulata* and

Figure 22. Semi-evergreen Intermontane *Acacia* Thorn-forest.
Relict patches in ravines. Note heavily eroded slopes in background and a fog bank moving up through the valley. Río Leones valley, below Oña, Azuay Province, Ecuador (1,850 m, 14 February 1991).

Chorisia sp., suggesting that the original vegetation type might have been a drier version of the *Ceiba trichistandra* forest, though none of the characteristic species have been found.

In Peru much of the western Andean slope south of 8°S is covered by quite similar cacti and scrub steppes and forests at altitudes between 1,000 and 3,500 m. Here too, human exploitation makes it very difficult to find natural-looking vegetation patches. Detailed descriptions have been given by Weberbauer (1945) and Koepcke (1961).

HUMID TO VERY HUMID LOWER MONTANE CLOUD-FOREST

(Natural) limits	**lower**	**upper**
elevation (m)	1,400-1,500	1,700-1,800
mean annual precipitation (mm)	1,300	3,000+

Climate This is another vegetation type with a short potential dry season which is completely prevented by the very frequent cloud condensation, which occurs mainly in the 'dry' season. Mean annual temperatures are about 18 to 20°C.

Occurring on the western slopes this forest type (together with the Humid Premontane Cloud-forest) represents the southernmost extension of a belt of humid montane rainforest stretching north to Colombia, Central America and over large areas of the eastern Andean slopes. Only small fragments are found further south into western Peru, e.g. at Hacienda Taulis (Koepcke 1961: 154-160, Dillon *et al.* 1993). Its species composition therefore shows more affinity to those northern areas than to the adjacent arid fringes.

The forest is 20 to 30 m tall, with a dense epiphyte-laden canopy (Figure 23). Trees reach 1 m dbh; strangler figs, palms and buttressing are all frequent. The undergrowth is rather open with such characteristic plants as *Chamaedorea* palms, *Anthurium*, and *Heliconius* and *Costus* along forest margins.

At Tierra Colorada trees are difficult to sample and the following list is very incomplete:

Ficus spp.
Cecropia sp.
Casearia cf. *sylvestris*
Erythrina sp.
Annonaceae
Dendropanax
Gustavia sp.
Meliaceae
Ochroma sp.
Inga spp.
Palms
Oreopanax sp.
Eschweilera sp.
Syzygium jambos
Cupania cf. *latifolia*
Rheedia cf. *edulis* (very common)
Lauraceae

Figure 23. Humid Lower Montane Cloud-forest.
Tierra Colorada, Loja Province, Ecuador (1,600 m, 12 February 1991).

Understorey species include:

Miconia spp.
Leandra cf. *longicoma*
Geissanthus sp.
Carica microcarpa
Myriocarpa sp.
Allophylus cf. *scorbiculatus*
Clavija repanda
Chamaedorea pinnatifrons
Tournefortia sp.
Verbesina sp.
Cordia sp.
Citharexylum sp.
Cyathea spp.
Aphelandra sp.
Siparuna sp.
Mollinedia sp. nov.

Vines, which are only common along forest edges, include:

Marcgravia cf. *coriacea*
Anemopaegma puberulum
Amphiliphium paniculatum
Paullina dasytachya
P. cf. *exalata*
Dioscorea spp.
Passiflora spp.
Smilax sp.

Epiphytes are fairly common, but do not occur in dense mats or as 'beards' as they do a few hundred metres higher up on these mountains:

Peperomia spp.
Comparettia falcata
Masdevallia spp.
Araceae
Polypodium spp.
Campyloneurum sp.
Gesneriaceae

Bromeliads include:

Aechmea fraseri
Guzmania fusispica
Tillandsia complanata
T. floribunda
T. hamaleana
T. narthecioides
T. umbellata

DECIDUOUS TO SEMI-EVERGREEN LOWER MONTANE CLOUD-FOREST

(Natural) limits	**lower**	**upper**
elevation (m)	1,300-1,400	1,800-2,000+
mean annual precipitation (mm)	400-800	900-1,300

Climate Lying on the upper edge of mountain ranges this vegetation type can be affected by cloud condensation to varying degrees. The dry season is four to six months long, mean annual temperature is in the range of 17 to 20°C.

This vegetation type might well be divided into two, one with and one without strong cloud influence. However, the only patch of forest without strong mist influence studied (Catacocha) had a very similar species composition to the cloud-forest near Sozoranga. It differed mainly in having fewer epiphytes. Areas between 1,000 m (the upper limit of *Ceiba trichistandra* forest) and 1,450 m (the lower edge of the lowest forest patch found) are so densely inhabited, that not a single forest patch could be located. Therefore the transition from the *Ceiba trichistandra* forest to this forest type could not be studied.

This very characteristic forest type has a restricted distribution along the rim of the mountain ranges surrounding the lower Catamayo and Catacocha valleys. It receives only limited precipitation but is often shrouded in mist. Many trees are drought-deciduous and others commonly have small, sclerophyllous leaves. Along damp ravines evergreen gallery forest is found. Drought-resistant epiphytes (mainly 'grey' bromeliads and cacti) are abundant, receiving sufficient moisture throughout the year (Ellenberg 1975)(Figures 24 and 25).

Trees reach 10 to 15 m, in humid ravines 20 m. Trunks can reach 0.6 dbh. and buttressing is rare. Steepness of the terrain and partial deciduousness of the trees give rise to an irregular canopy which permits the development of a very dense middle-storey, composed mainly of immature specimens of the canopy-forming species. Even in the rainy season herbs are virtually absent within the closed forest.

Trees recorded around Sozoranga and Catacocha include:

Jacaranda sparrei
Tabebuia chrysantha
Erythrina sp. *(edulis* and/or *triana)*
Acacia macracantha
Oreopanax sp. in humid ravines
Sapindus saponaria
Triplaris cumingiana
Schmarckea microphylla
Chorisia insignis
Clusia sp. in humid ravines
Inga sp.
Ficus spp. including strangling species
Ficus aff. *cuatrecaseana*

Figure 24. Deciduous to Semi-evergreen Lower Montane Cloud-forest.
Relict forest patch below Catacocha, Loja Province, Ecuador. *Fourcroya* (Agavaceae) is growing on steep cliffs. Note cleared area above forest and field in lower left corner. (1,500 m, 7 March 1991).

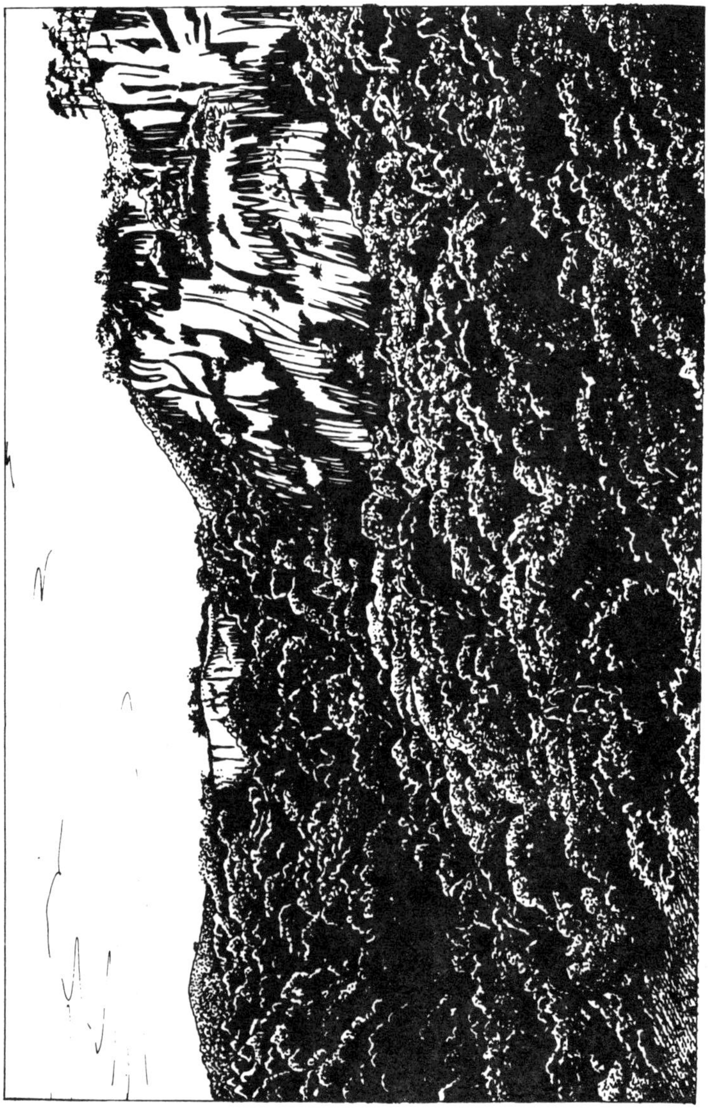

Figure 25. Deciduous to Semi-evergreen Lower Montane Cloud-forest.
Trees covered with bromeliads. Quebrada Suquinda, W of Sozoranga, Loja Province, Ecuador (1,600 m, 30/31 January 1991).

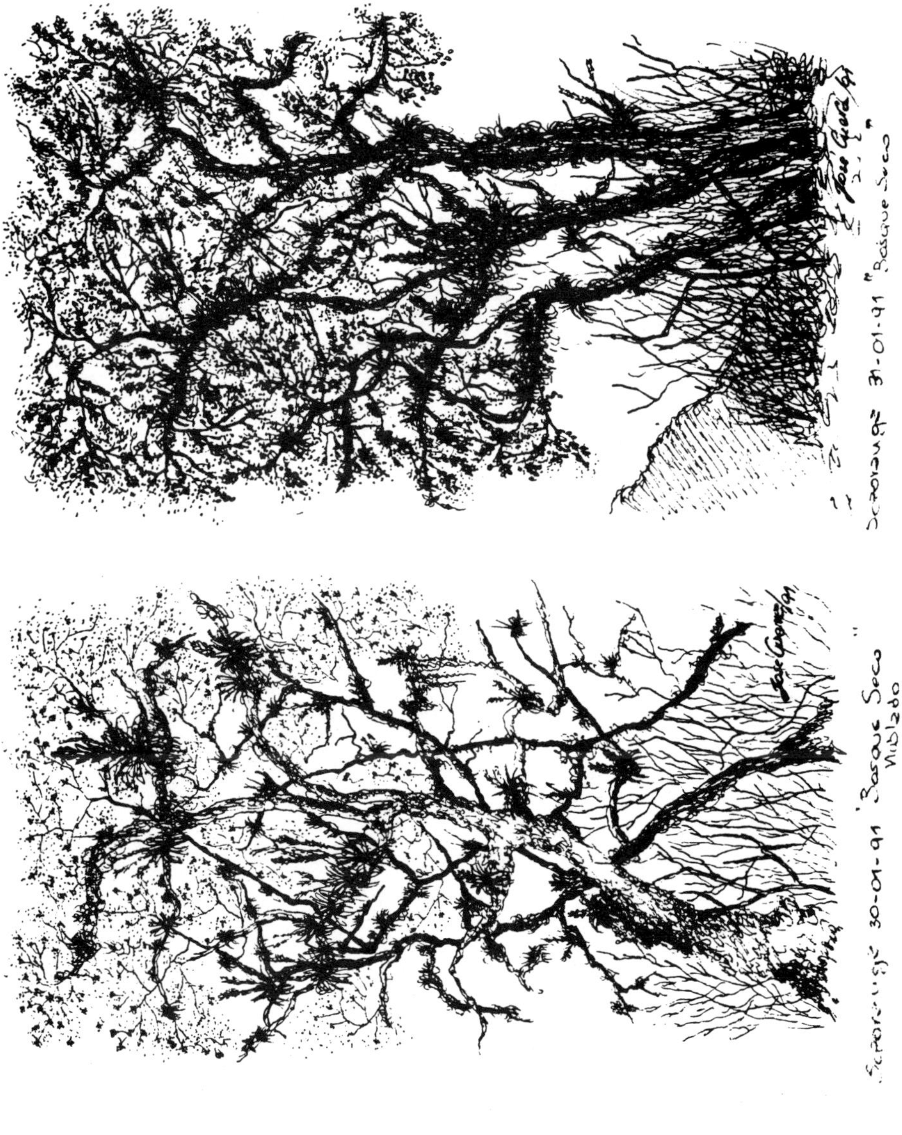

Middle-storey species:

Acalypha sp.
Solanum spp.
Croton spp.
Chlorophora tinctoria
Cornutia sp.
Coccoloba ruiziana
Clavija euerganea
Rubiaceae in humid ravines
Chusquea sp.

Among the few understorey herbs the following are found:

Oxalis spp.
Gesneriaceae
Polypodium sp.
Adiantum sp.
Begonia spp.

The abundant epiphytes include:

Peperomia sp.
Araceae
several ferns (*Polypodium* spp. and others)
Gesneriaceae
Loranthaceae

and the bromeliads:

Guzmania monostachya
G. variegata
Pitcairnia cf. *oblongifolia*
Tillandsia disticha
T. flagellata
T. floribunda
T. multiflora
T. polyantha
T. pugiformis
T. sagasteguii
T. sinuosa
T. straminea
T. triglochinoides
T. usneoides (rare)

HUMID TO VERY HUMID MONTANE CLOUD-FOREST

(Natural) limits	**lower**	**upper**
elevation (m)	1,700-2,000+	?
mean annual precipitation (mm)	1,000	2,500+

Climate Only occurring on the highest, most exposed mountain ranges this vegetation type has the lowest mean annual temperatures (about 16 to 18°C) and a constantly humid climate strongly influenced by the very frequent cloud banks hanging on the mountain slopes.

This forest type is found throughout south-west Ecuador above 2,000 m, except in the very arid upper Jubones valley and the mountains near Cariamanga. Whilst found only locally below 2,000 m it is familiar from many other locations and studies throughout Ecuador (e.g. the Podocarpus National Park); this vegetation type was therefore visited only briefly for this study.

The examples of this forest type studied were often relatively dry in comparison with other sites, however the species composition was very typical (if somewhat depauperate) and the appearance of trees equally characteristic, with twisted trunks and branches and dense, globular clumps of foliage (see Troll 1968).

Large trees can reach 15 or even 20 m, often with a diameter of over 1 m. Undergrowth is often dense and dominated by *Chusquea* bamboo. Epiphytes (mosses, lichens and kormophytes) cover all available surfaces and often form dense 'beards' (liverworts of the family Meterioriaceae) hanging down from branches (Figure 26).

Species found at Celica and above Sozoranga are:

Trees:

Clusia spp.
Weinmannia sp.
Hesperomeles ferruginea
Erythrina sp.
Cupania cf. *latifolia*
Juglans neotropica
Oreopanax sp.
Inga spp.
Myrtaceae

Understorey trees and shrubs:

Rubiaceae very common
Myrtaceae very common
Miconia cf. *serrulata*
M. lutescens
Siparuna macrophylla
Styrax sp.
Iochroma sp.
Acalypha sp.
Chusquea common
Piper spp.
Monnina sp.
Rosa sp.

Figure 26. Humid Montane Cloud-forest.
Low, dense, moss-covered forest on ridge-top above Sozoranga, Loja Province, Ecuador (1,850 m, 1 February 1991).

Rubus sp.
Berberis sp.
Tibouchina laxa
Liabum sp.
Baccharis sp.

The latter seven genera were particularly prominent along the forest edge.

Bromeliads include:

Guzmania variegata
Pitcairnia cf. *prolifera*
Tillandsia confertiflora
T. floribunda
T. cf. *fraseri*
T. cf. *pallidoflavens*
T. cf. *pastensis*
T. pyramidata
T. somnians

Further vegetation types of the Tumbesian region

The vegetation of the entire Tumbesian region is shown in Figure 27. Those vegetation types not found in the provinces of El Oro and Loja (and therefore not discussed in detail above) are briefly described below. The map and descriptions are based on data from Weberbauer (1945), Koepcke (1961), Tosi (1968), Gentry (1977), Cañadas Cruz and Estrada (1978), Harling (1979), Valverde *et al.* (1979), Valverde (1991), C. Josse (verbally 1992), Parker and Carr (1992), and personal observations made in 1984 and 1986 in Tumbes and Piura Departments, Peru, and in 1991 in Loja, El Oro and Azuay Provinces, Ecuador. Some general trends are easily discernible from the map:

- From the Peruvian border southwards aridity increases markedly, restricting the more mesic vegetation types to the foothills. The wet forest types reach south to 7°S on the Andean slopes, while the *Ceiba* forest, which only goes up to 1,600 m, disappears at 5°30'S. Fragments of thorn-forest are found south to 10°S in arid valleys in central Peru.
- Similarly, vegetation types occur at higher altitudes in the southern part of their distribution.
- The provinces of Loja and El Oro show the highest diversity of vegetation types. This is caused by the complicated topography of the region.

Desert

Areas with less than 100 mm annual precipitation are almost devoid of vegetation. Typically only a few bromeliads (*Tillandsia*) and cacti, as well as plants which survive the dry season as seeds (annuals, therophytes) or with bulbs (geophytes) are found.

Figure 27. Vegetation map of the Tumbesian region.
See text for descriptions of vegetation types.

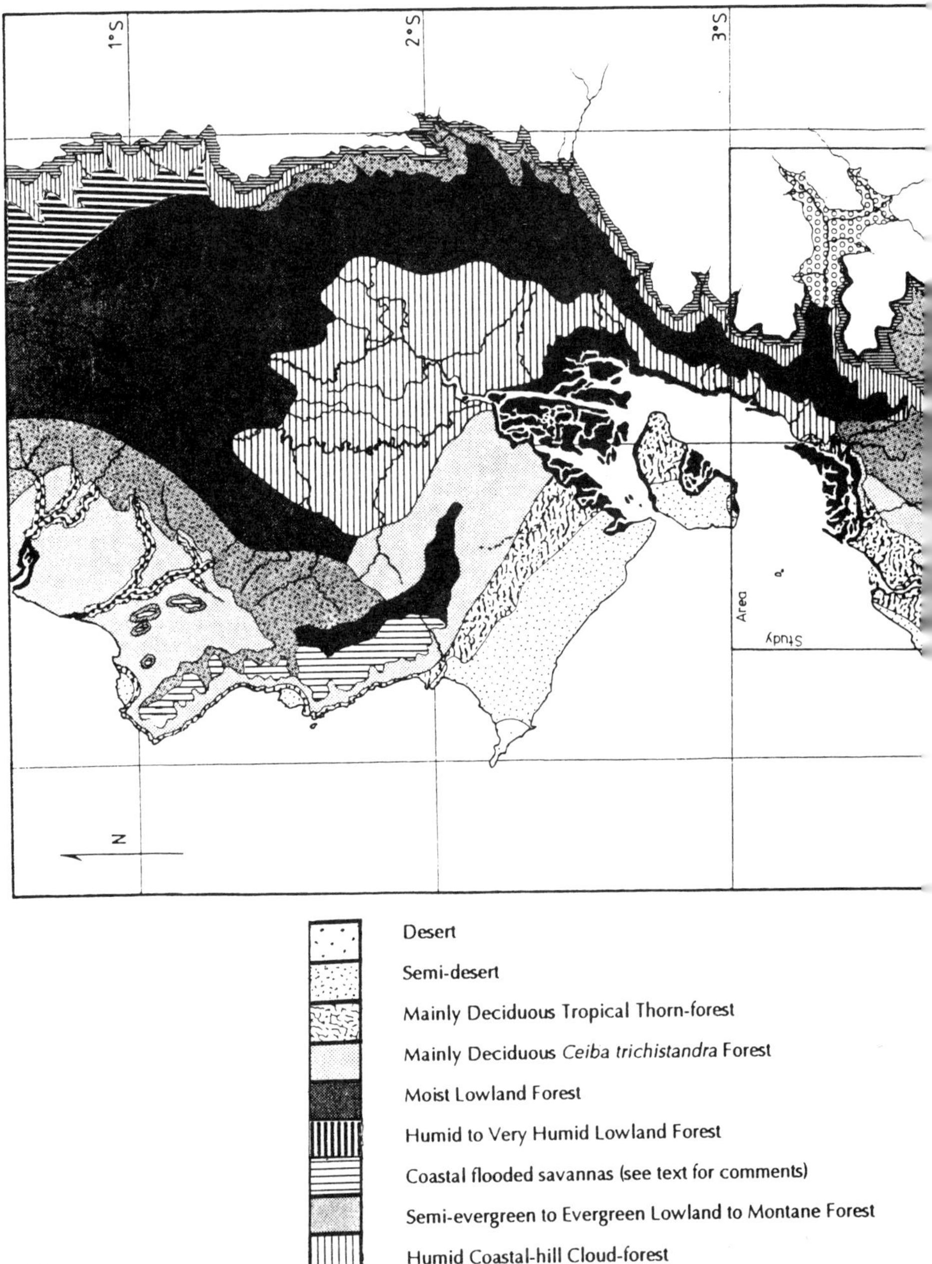

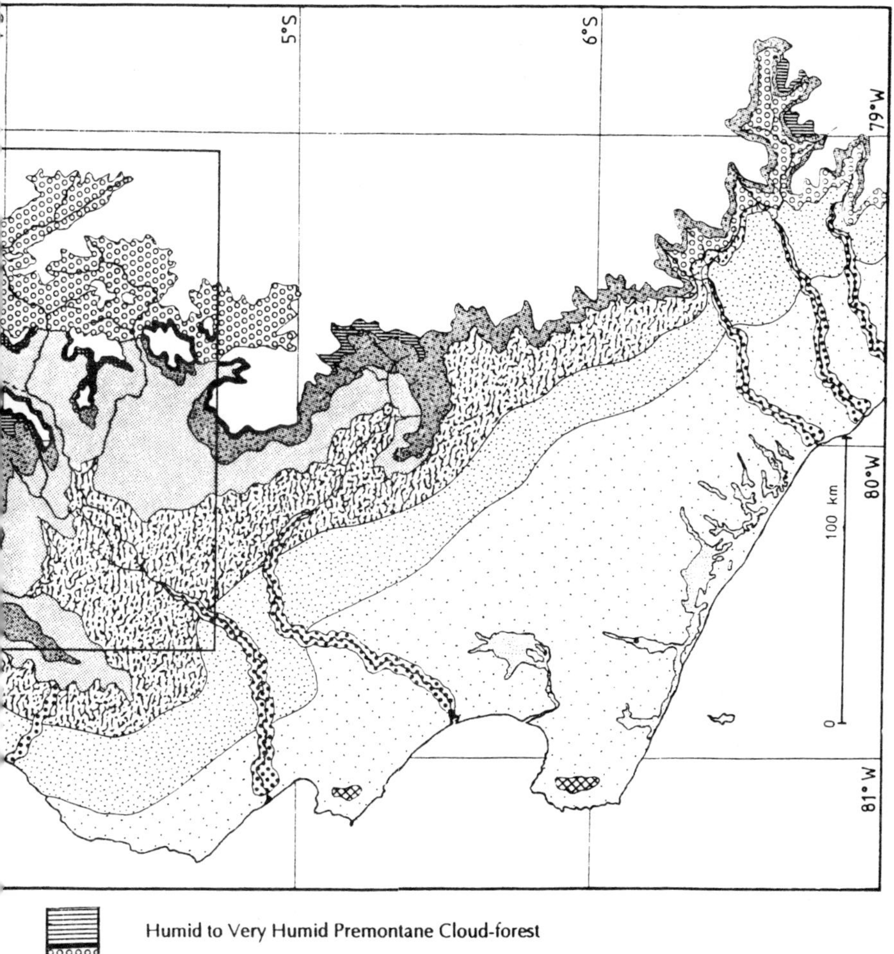

Humid to Very Humid Premontane Cloud-forest

Deciduous to Semi-evergreen Intermontane scrub, Thorn-forest and Forest

Humid Lower Montane Cloud-forest

Deciduous to Semi-evergreen Lower Montane Cloud-forest

Humid to Very Humid Montane Cloud-forest

Mangrove Forest

Salitrales

Riverine Forests

Lomas

Semi-desert

Those lowland areas with no more than 250 mm precipitation per year are covered in thorny, mostly deciduous scrub. In the rainy season numerous ephemeral and geophytic plant species impart an aspect of lushness. Characteristic species are *Cercidium praecox* and *Capparis* spp., and on the northern Peruvian foothills cacti-steppes of *Neoraimondia gigantea, Armatocerues cartwrightianus* and *Haageocerus* spp. (Koepcke 1961). Extensive use as grazing grounds for cattle and goats has led to the degradation of all examples of this vegetation type. More detailed descriptions are provided by Weberbauer (1945), Koepcke (1961) and Valverde *et al.* (1979).

Humid Coastal-hill Cloud-forest

The coastal hills of the provinces of Guayas and Manabí reach some 850 m in altitude and can therefore intercept a considerable amount of rain and fog (Troll 1968). The vegetation of the area is described in Valverde (1991) and Parker and Carr (1992). While the lower parts of the hill range are covered in semi-evergreen thorn-, *Ceiba trichistandra* and moist lowland forest, the upper slopes (above 600m) support a peculiar type of cloud-forest which is most similar to the moist and wet forests further north and east along the Andean foothills. Until a detailed comparison has been made, and given the geographical isolation of the area, it seems reasonable to consider this a different vegetation type. For the deciduous *Ceiba trichistandra-*, *Ceiba pentandra-* and Humid Coastal-hill Cloud-forests of the Chongón-Colonche Cordillera, Valverde (1991) lists 671 plant species including 174 trees, 126 shrub species, 204 herbs, 125 lianas and vines, 42 epiphytes and 9 parasitic and hemiparasitic species. Some characteristic trees of the more humid forests include (Valverde 1991, C. Josse verbally 1992):

Quararibea grandiflora
Calophyllum sp.
Grias peruviana
Cynometra sp.
Ocotea spp.
Guarea glabra and *pterorhachis*
Inga spp.
Cecropia obtusifolia
Clarisia cf. *racemosa*
Coussapoa villosa
Poulsenia armata
Pseudolmedia rigida
Triplaris cumingiana
Sparrea schippii
Simira ecuadoriensis
Sapium utile
Carapa guianensis

Stands of the huge bamboo *Bambusa guadua* give the forests a very characteristic appearance. Of the 15 bromeliad species listed, three (*Aechmea mexicana, A.*

angustifolia, and *Tillandsia monadelfa*) have not been recorded in the provinces of Loja, El Oro and Azuay, indicating that there is a certain floristic differentiation between the humid hill forests of the Coastal Cordillera and those of the Andean foothills. These three species are all widespread in the humid forests of Central America and east of the Andes. Parker and Carr (1992) further provide extensive descriptions and species lists from several sites along the Coastal Cordillera.

Semi-evergreen to Evergreen Lowland to Montane Forest

This vegetation type combines the Semi-evergreen *Ceiba pentandra* forest and the Semi-evergreen Lowland to Premontane Tall Forest described for Loja and El Oro and, additionally, includes evergreen forests and, in Peru, montane forests. Therefore, this vegetation type forms the transition between dry, deciduous forest types and wet, evergreen forest types. While floristic composition and forest structure are certainly quite different on the coastal hills in Manabí, Ecuador, and in the forest belt between 1,200 and 2,000 m in Lambayeque, Peru (Koepcke 1961) at the present state of knowledge it is not possible to differentiate this vegetation type any further (as was done above for south-west Ecuador).

Coastal flooded 'Savannas' of Guayas and El Oro Provinces

Much of the coastal lowlands of El Oro Province and areas within the lower reaches of the rivers Daule and Babahoyo in Guayas Province are flooded annually. Some authors (e.g. Acosta-Solis 1968, Troll 1968) describe these areas as grass-dominated savannas with gallery forests and occasional clumps of trees, while others (e.g. Dodson and Gentry 1991) treat them as dry forest areas. Interestingly, Dodson and Gentry (1991) map these areas as devoid of forest on a map of the aboriginal forest cover of western Ecuador but give no explanation. Today, extensive monocultures and cultivation have left practically no natural vegetation, and as long as no detailed study is made of the vegetation history, it will not be possible to define the potential natural vegetation of the region and the extent of past human influence.

Azonal vegetation types

Azonal vegetation types are those whose occurrence is determined by very special, local climatic or soil conditions, and therefore are not primarily affected by the general climatic conditions. Four characteristic azonal vegetation types are found in the Tumbesian region; since they cover only small areas and do not belong to the Tumbesian dry forests as such, they will be mentioned only briefly.

<u>Coastal Mangrove Forest</u>

A narrow strip of mangrove forest follows much of the coast south to about 3°S. In some shallow and estuarine areas of islands and shifting tidal inlets larger areas can be covered (Figure 27). The constituent species are *Rhizophora mangle* (*R.*

harrisonii according to Clusener and Breckle 1987), *Avicennia germinas, Laguncularia racemosa* and *Conocarpus erectus* (Acosta-Solis 1959, de Macedo 1979, Vareschi 1980, Jordan 1988). Usually *Rhizophora* grows in the deepest water, followed further inland by *Laguncularia*, the *Avicennia* and finally *Conocarpus* (Jordan 1988). Islands typically have a belt of mangrove forest along the edges and xeric woodland further inland which is composed of such tree species as *Prosopis*, *Acacia macracantha, Scutia spicata, Baccharis lanceolata* and *B. salicifolia* (de Macedo 1979).

Salitrales

Locally in depressions along the coast and in many places inland of the mangrove belt, salt evaporation pans, so called 'salitrales', may be found. Often they may contain such species as *Sesuvium portulacastrum, Salicornia fruticosa, Batis maritima, Cressa truxillensis* and the grass *Distichlis spicata* (Koepcke 1961, Jordan 1988).

Riverine forests

Throughout tropical South America the vegetation succession along rivers is very similar and was observed in south-west Ecuador along the río Sabiango at Tambo Negro in Loja Province (Figure 28). Fresh sand or mud banks along rivers are colonized by numerous Cyperaceae, Juncaceae, grasses and sedges, *Ludwigia, Cuphea, Cleome, Plumbago scadens* and young *Tessaria integrifolia*. Areas which are not flooded during the annual rainy season are dominated by thickets of *Tessaria* up to 2 m high.

These thickets are gradually colonized by small trees e.g. *Salix humboldtiana, Alnus acuminata, Acacia macracantha, Piper* sp., *Desmanthus* sp. and the tall grass *Gynerium sagittatum*. Further away from the river in areas not subject to flooding but where high ground-water levels occur, a more species-rich gallery forest may develop. Its species composition is markedly influenced by the surrounding vegetation. Species found at Tambo Negro include *Hura crepitans, Muntingia calabura,* mature *Salix humboldtiana, Prosopis* sp., and numerous dry forest components such as *Cochlospermum vitifolium* and *Erythrina velutina*.

In the northern Peruvian Sechura Desert such riverine forests represent the only evergreen vegetation and are vital for the survival of many animal species. In areas with groundwater, forests of *Prosopis* may be found several kilometres away from rivers (Weberbauer 1945, Koepcke 1961). Other characteristic species are *Capparis angulata, Celtis iguanea, Caesalpinia corymbosa, Parkinsònia aculeata* and *Zizyphus piurensis*.

In the dry forest area of Manabí Province, Ecuador, Troll (1968) maps evergreen gallery forests, but gives no descriptions or species lists.

Figure 28. Riverine forest along the río Sabiango at Tambo Negro.
1: Flooded river-bed a few days after the onset of the annual rains. 2: *Tessaria/Salix* thickets. 3. High-ground forest with access to ground-water. 4. Mainly deciduous *Ceiba trichistandra* forest on higher slopes. The tall, dark-trunked trees are *Ceiba*.

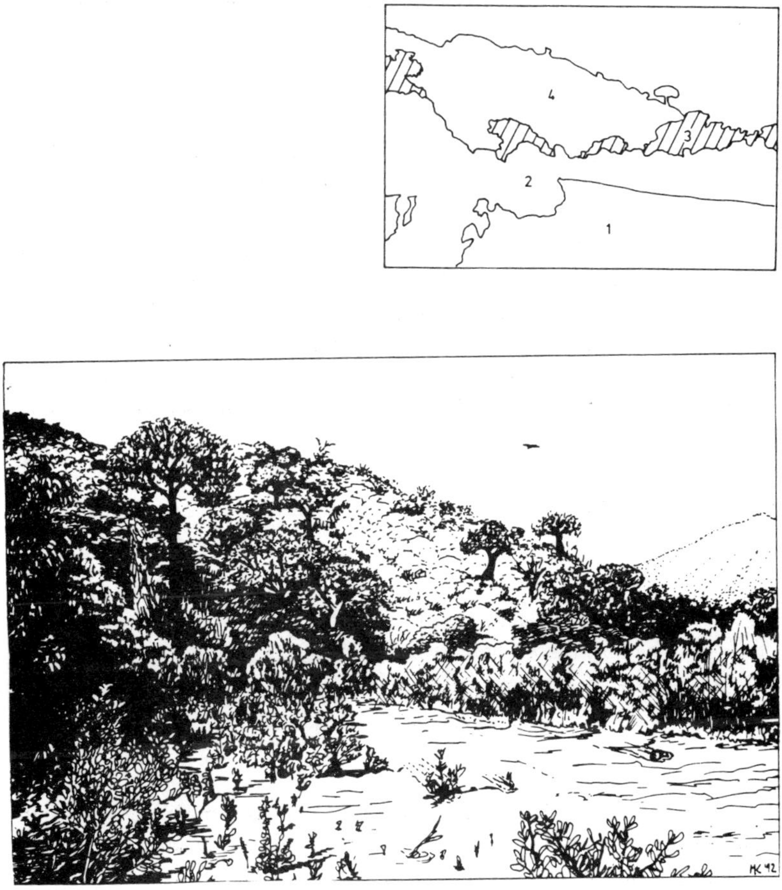

Lomas

This vegetation type, which is typical of the Peruvian and Chilean coastal desert, is formed on coastal hills by frequent winter fog banks which are almost the only source of humidity in areas which otherwise receive only a few millimetres of rainfall a year. The northernmost *lomas* are found on Cerro Illescas and Silla de Paita in Piura Department, Peru (Koepcke 1961), and have been described by Weberbauer (1945).

ECOLOGICAL FACTORS AND THE DISTRIBUTION OF VEGETATION TYPES

The distribution of the 10 zonal vegetation types in south-west Ecuador is shown in Figures 29, 30, and 31; Figure 27 showed the vegetation of the entire Tumbesian region. In order to correlate environmental factors with the distribution of particular vegetation types in south-west Ecuador, Figure 32 shows the distribution of the vegetation types in that area as determined by precipitation and elevation.

In south-west Ecuador the climatic factors most influencing the vegetation are rainfall and humidity. Generally speaking there are four important patterns:

- Precipitation decreases from north to south.
- Precipitation decreases further inland as a result of the rain-shadowing effect of the mountain ranges.
- With increasing altitude and correlated decreases in temperature and evapotranspiration, humidity increases.
- Seaward-oriented mountain ranges often receive additional moisture from fog condensation, especially in the dry season.

As might be expected, these general trends are variously modified by topographic factors; furthermore they interact, so creating an extremely complicated pattern of different bioclimates which can change over very short distances. The transition from evergreen tropical rainforests near Pasaje, El Oro Province to the xeric cactus scrub in the central Jubones valley, which can be witnessed in a 30 minute car drive along the Machala-Cuenca road, represents a dramatic example of this high local habitat diversity. There are probably few areas in the world where such extreme climatic and vegetational variations can be found.

Figure 29. Distribution of vegetation types of south-west Ecuador below 2,000 m.

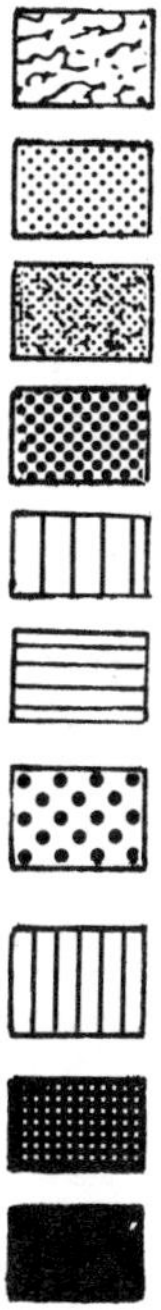

- Mainly Deciduous Tropical Thorn-forest and *Acacia* Thorn-forest
- Mainly Deciduous *Ceiba trichistandra* Forest
- Semi-evergreen *Ceiba pentandra* Forest
- Semi-evergreen Lowland and Premontane Tall Forest
- Moist Lowland Forest
- Humid to Very Humid Premontane Cloud-forest
- Deciduous to Semi-evergreen Intermontane Scrub, Thorn-forest and Forest
- Humid Lower Montane Cloud-forest
- Deciduous to Semi-evergreen Lower Montane Cloud-forest
- Humid to Very Humid Montane Cloud-forest

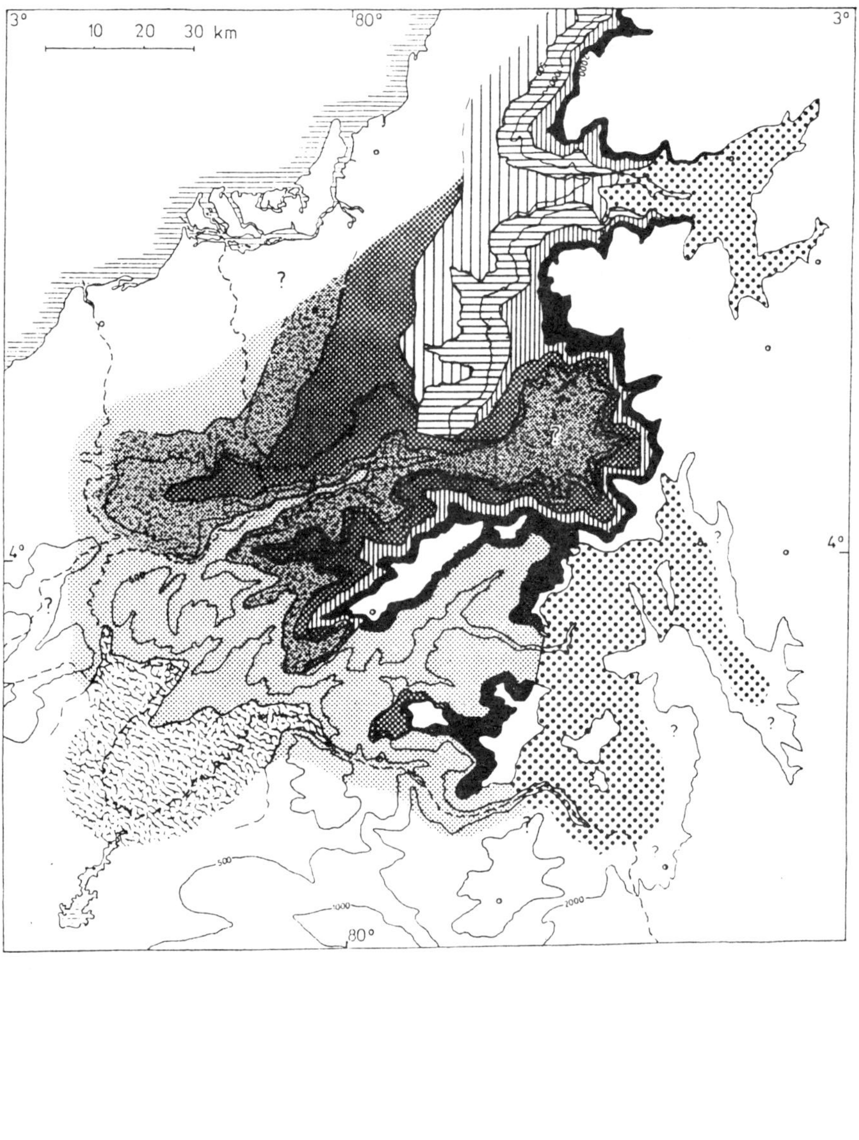
3°
10 20 30 km
80°
3°
4°
4°
500
1000
2000
80°

The formation of fog banks influences the vegetation to a surprising degree. Thus the coastal lowlands (e.g. around Arenillas, El Oro Province) which regularly receive additional moisture from mist, support high, semi-evergreen *Ceiba trichistandra* forests, while further inland between Sabanilla and Zapotillo in an area with similar precipitation but considerably higher insolation, only low, open, mainly deciduous thorn-forests are found. The effect of fog condensation and shading by clouds in coastal areas can apparently be equivalent to up to 500 mm precipitation. This explains why in Figure 31 the lowest areas with least rainfall (coastal areas) have more mesic vegetation types than higher areas with more precipitation (further inland). This effect becomes less evident as precipitation increases.

A special combination of conditions exists in northern El Oro and southern and central Azuay Provinces. Here the western Andean slope rises steeply from the narrow coastal plain (only 20-30 km wide) which separates the mountains from the Gulf of Guayaquil, concentrating the precipitation in a narrow area of foothills and slopes, thus making them locally the wettest area in the region. Another factor might be the proximity to the warm waters of the Gulf of Guayaquil which probably lead to the regular formation of fog banks, especially in the dry season when relatively little rainfall falls. This compensates much of the potential aridity of the dry season, thus creating a truly evergreen forest area.

Further south the generally lower precipitation falls over an area of several successively higher mountain ranges, while further north much of the rain falls on the low coastal hill range and on the wide (175 km) area of lowlands before reaching the Andes. All these factors contribute to the formation of a more arid gap between the wet premontane forests of El Oro and Azuay Province and those further north in Bolivar Province. This pattern is not clearly discernible on available maps which are based on the Holdridge system and on scanty climatic data (Cañadas Cruz and Estrada 1978, Dodson and Gentry 1991).

As delimited here, vegetation types are influenced primarily by climatic factors. However, within a vegetation type, different geological substrates and soils may lead to changes in species composition. Thus, on different soils physiognomically similar forests may contain completely different species assemblages. In Amazonia, Gentry (1988) found little overlap: less than 20% in tree species composition between forests on different soil types, and concluded that much of the diversity of woody plants in this region as compared to the rest of the Neotropics is largely due to habitat specialization. Whether such conclusions may also apply to the Tumbesian region remains to be studied, but in any case edaphic differentiation would take place within the vegetation types as defined here, and is not responsible for the differentiation of the vegetation types. It might be assumed that edaphic factors play a more important role in vegetation types with a permanently humid climate, because under humid climatic conditions soil development is faster than under arid conditions and because in humid localities plants are not forced to adapt to climatic environmental stress, but rather react to different soil conditions.

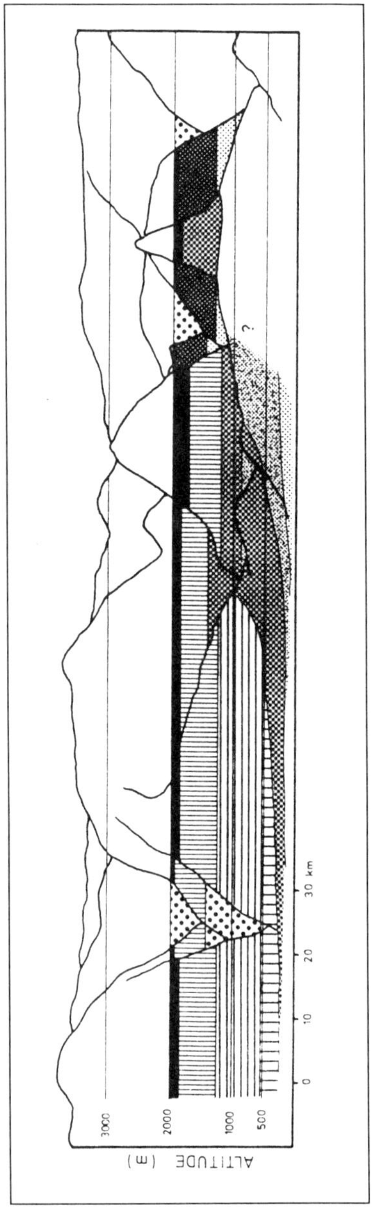

Figure 30. Altitudinal distribution of vegetation types of south-west Ecuador as seen from the west. Signatures as in Figure 29.

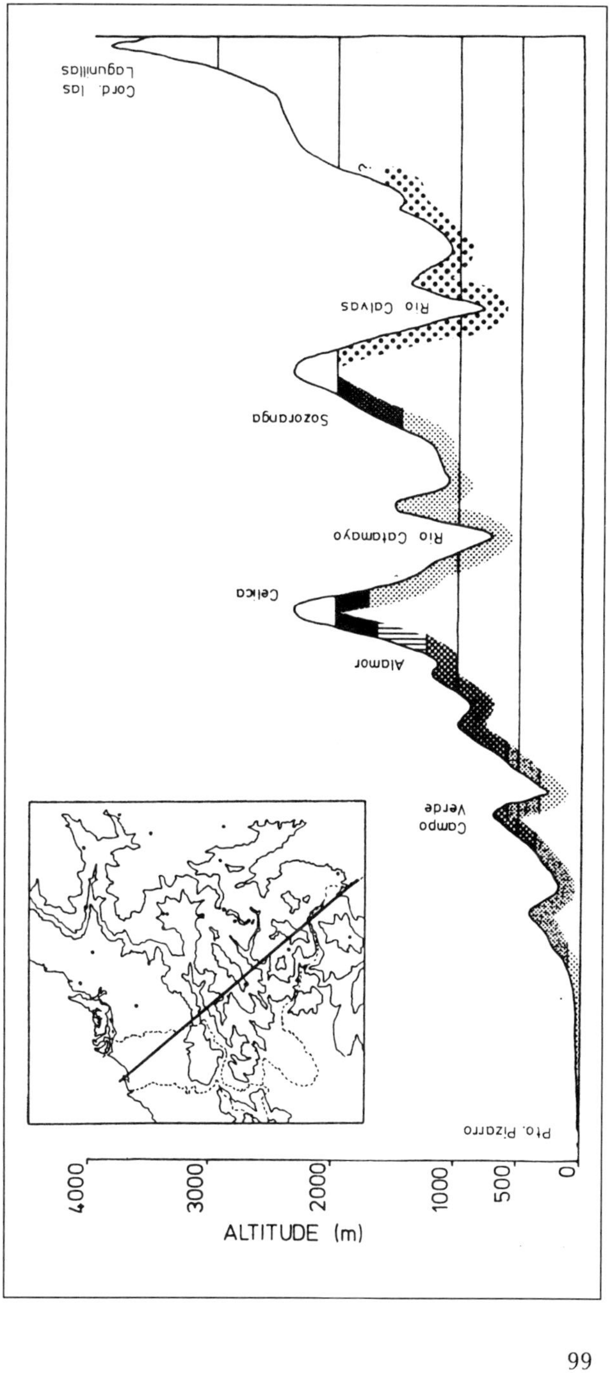

Figure 31. Cross-section through the western Andean slope in south-west Ecuador showing distribution of vegetation types. Signatures as in Figure 29.

Figure 32. Schematic distribution of south-west Ecuadorian vegetation types as determined by altitude and precipitation. Row of white circles delimits region with regular occurrence of mist and fog, row of black dots region of very frequent occurrence of mist and fog. See text for additional comments.

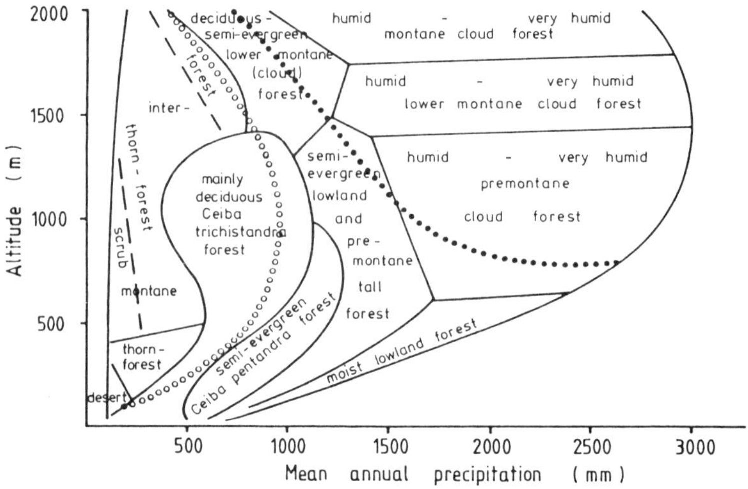

COMPARISON WITH OTHER CLASSIFICATION SYSTEMS

The phytogeographical classification presented here has been compared to those of Holdridge (1956), Acosta-Solis (1968) and Harling (1979). Certainly the most important of these is the Holdridge school, as it has been used by Tosi (1960) in Peru, and Cañadas Cruz and Estrada (1978) and Dodson and Gentry (1991) in Ecuador.

For a direct comparison, in Figure 33 the Holdridge system has been adapted to the same form of presentation as the ecogram in Figure 32. As can be seen, the Holdridge system is unsatisfactory in several ways:

- It only shows two altitudinal belts for south-west Ecuador.
- Since it is based only on rainfall data (the potential evapotranspiration used

Figure 33. Classification system of Holdridge (1956) as adapted to the same form of presentation as Figure 32. See text for comments.

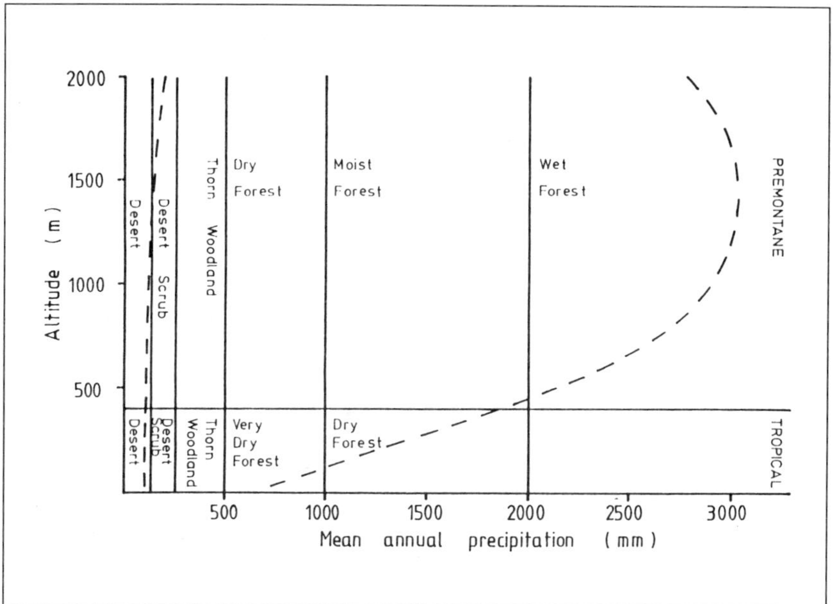

in the system is calculated from precipitation and temperature) it completely misses the fog-effect so conspicuous in the region. Thus is an area which (based on the Holdridge system) should have little more than Thorn Woodland, one can actually find 30-m-high *Ceiba* forests. As shown by Borhidi (1976) Holdridge's potential evapotranspiration constant is only valid for tropical lowland climates which are wet all the year.

- The terminology is defined by climatological data, but implies physiognomical characteristics (e.g. Thorn Woodland), and these have not been defined (Borhidi 1991).

As pointed out by Acosta-Solis (1977:296) the Holdridge system should only be used as a guide to what vegetation type might theoretically be found in an area, and if a discrepancy is found, this might help to discover why a region has a more mesic or arid vegetation. Unfortunately the system has been uncritically applied to many Latin American and some African and Asian countries and, rather than helping our understanding of complicated vegetation patterns of those areas, this has created the illusion of a complete and definitive coverage and has slowed the gathering of knowledge on the vegetation in these regions.

The classification by Acosta-Solis (1968) does not even include many of the encountered climatic conditions (e.g. it does not describe vegetation types between 800 and 2,000 m with precipitation less than 1,000 mm a year) and is therefore of little use in south-west Ecuador.

Harling's (1979) classification of the Ecuadorian vegetation lists eight vegetation types as occurring in south-west Ecuador, but consistently interprets the vegetation to be more arid than is actually the case. *Ceiba* forests, for example, are treated as "savanna". Disturbed *Ceiba* forests can certainly look like savanna, but when undisturbed they form tall forests with high cover values in the rainy season. In a similar way all of the intermontane region of Loja Province is described as inter-Andean desert, semi-desert and dry scrub. No mention is made of the forests which originally covered large areas of the inter-Andean valleys, of which relict patches can be found throughout the region today. In general it can be said that Harling's (1979) view is strongly influenced by the anthropogenic destruction of most of the vegetation.

SPECIES DIVERSITY AND ENDEMICITY

It is very difficult to give accurate estimates of diversity and endemicity in the scope of such a brief study. Difficulties include:

- Some vegetation types were not visited at all, and those that were necessarily insufficiently sampled.
- Many important and species-rich families are very poorly known in Ecuador e.g. Compositae, Solanaceae, Bromeliaceae and Graminaceae (G. Harling *in litt.* 1991).
- The flora of the south-western part of Ecuador is particularly poorly known.

The following analysis concentrates on the province of El Oro, the province of Loja west of the continental water divide and Azuay north to Cuenca and Molleturo. A brief study of the *Flora of Ecuador* (Harling and Sparre, since 1979)(including 1,785 species from 33 families, but excluding Melastomataceae and Bromeliaceae) was undertaken to find general distributional patterns. Estimating a number of 20,000 plant species for Ecuador (Gentry 1978, Harling 1986, Jørgensen *et al.* 1993) this database covers less than 10% of the flora. Results should be interpreted with respective care. Of the 1,785 species covered in Harling and Sparre 542 (30%) are found in south-west Ecuador; 368 (21%) of these also occur in neighbouring countries, some only to adjacent Peru and 174 (9%) are endemic to Ecuador. Of these 174 Ecuadorian endemics 62 (3%) also occur outside the south-west, 112 (6%) are restricted to the south-west and 51 (3%) are known from only one locality (in most cases from just one collection).

Extrapolation shows an estimated 6,000 plant species for south-west Ecuador of which 1,200 are endemic to the area. Future fieldwork will alter this figure by finding new species for the area, larger ranges for some endemics and new endemics. Many of the south-western endemics are restricted to fragmented páramo patches and cloud-forests. For example 8 of the 15 species of the genus Lysiopomia (Lobeliaceae) known for Ecuador have very limited ranges restricted to the south-west.

For a more detailed study of the distribution patterns a choice was made to focus on the Melastomataceae and Bromeliaceae, two important and characteristic plant families which are easily recognized in the field and are therefore likely to be sampled fairly completely. Conclusions drawn from the resulting data are necessarily biased by the choice of those families and the limited sample size (two out of over 200 plant families).

Melastomataceae

This family is particularly species rich with about 600 species recorded so far from Ecuador of which only 440 are treated in Wurdack (1980), the main source of information for this study. Sixty-eight specimens were collected in the field, but they only represent a fraction of the species occurring in the area. Also, since many specimens were not in reproductive condition, identification to species level was often not possible. According to Wurdack (1980) 103 species of Melastomataceae occur in south-west Ecuador (Table 4). An attempt was made to assign the collection localities to particular vegetation types, but in some cases this was difficult, especially where two or more forest types meet. Thus, species classified as occurring in both semi-evergreen forest types (mainly from localities near Zaruma and Portovelo in El Oro Province) would probably be better listed under a more humid vegetation type.

In spite of the rather inadequate database some trends are visible:

- 86 species (83%) have only been found in one vegetation type, thus showing a rather high habitat specificity. Future fieldwork will undoubtedly alter this figure.
- Endemicity increases with altitude: the Moist Lowland Forest has no endemics, the Humid to Very Humid Premontane Cloud-forest 53% (12 of and the cloud-forest above 2,000 m 22 species), the Humid to Very Humid Lower Montane Cloud-forest 73% and the cloud-forest above 2,000 m, 79%. The low figure of 50% for the Humid to Very Humid Montane Cloud -forest (below 2,000 m) is certainly due to its small distribution in the region and inadequate sampling.

Table 4. Occurrence of 113 species of Melastomataceae in seven vegetation types in south-west Ecuador.

Forest type	Distribution				
	Widespread	Ecuador endemic	SW endemic	Local endemic	Total
Semi-evergreen Lowland & Premontane Tall Forest	4(3)	-	1(1)	-	5(4)
Deciduous to Semi-evergreen Lower Montane Cloud-forest	1(1)	1(0)	1(0)	-	3(3)
Moist Lowland Forest	4(4)	-	-	-	4(4)
Humid to Very Humid Premontane Cloud-forest	10(7)	7(3)	4(0)	1(1)	22(11)
Humid to Very Humid Lower Montane Cloud-forest	4(1)	5(1)	6(2)	-	15(4)
Humid to Very Humid Montane Cloud-forest	4(4)	1(1)	2(1)	1(1)	8(8)
Cloud-forest above 2,000 m	12(11)*	7(7)	22(21)	16(16)	57(55)
Total	**39(31)**	**20(11)**	**36(26)**	**18(18)**	**103**

Species are classified as **widespread** (also occurring outside Ecuador), **Ecuadorian endemics** also occurring in the south-west, **south-western endemics** and very **local** endemics (known from only one locality). Numbers in parentheses indicate the number of species restricted to a particular vegetation type. * denotes 6 species which only range to Peru. Data from Wurdack (1980).

- Local endemicity is almost restricted to humid areas above 2,000 m: 81% of the south-western endemics and 89% of the local endemics are found there.

This family has its highest diversity in the cloud-forests and páramos. In fact many of the local endemics have been found in the mountains to the east and south-east of the city of Loja, an area now protected in the Podocarpus National Park. The distributional pattern shown by the Melastomataceae is also to be expected in other families adapted to moist, high altitude conditions, such as Tropaeolaceae, Ericaceae, Lobeliaceae, Campanulaceae, Araliaceae, Lycopodiaceae, Selan-aginellaceae, Orchidaceae and others.

Bromeliaceae

This is another very diverse family which is relatively little known in Ecuador (G. Harling *in litt.* 1991). Since many species are epiphytic, this group is of particular interest as edaphic variables are of limited importance to the plants and therefore climatic variables can be assumed to be of primary importance for their distribution.

Bromeliads are probably good indicators of bioclimatic conditions and may be useful for the classification of vegetation types (Dodson and Gentry 1991, Richter 1991).

One hundred and ten specimens were collected in 1991 (Kessler 1992), representing at least 56 species (identifications by H. E. Luther, Sarasota); they are listed under the respective vegetation types. A detailed analysis will be presented in Kessler (in prep.). Further data were obtained from Gilmartin (1972), Smith and Downes (1974-1979), Ehlers and Ehlers (1990) and Luther *in litt.* (1991). Particular attention was paid to the genus Tillandsia.

To date about 125 bromeliad species have been recorded from south-west Ecuador, including 72 (50%) of the 144 *Tillandsia* species found in Ecuador. Of the 45 tillandsias endemic to Ecuador, 23 (51%) have been found in the south-west and 12 (27%) are restricted to it. The distribution and ecological amplitudes of these locally endemic species are shown in Figures 34 and 35. There appears to be no concentration of species at any particular location, but ecologically some differentiation becomes apparent: six (50%) species are found above 2,000 m, and nine (75%) species are restricted to fairly dry habitats with no more than 1,000 mm mean annual precipitation. The first pattern might be explained by the more fragmented distribution of vegetation types at higher elevations, while the second is probably an indication of the general tendency of tillandsias to occupy more arid areas than other bromeliad genera, e.g. *Aechmea* and *Guzmania*. The genus *Puya* is characteristic of high-elevation páramos and rocky outcrops, and has several endemic species in Azuay and Loja.

To test the usefulness of the vegetation classification proposed in this chapter, the ecological distribution of Tillandsia was plotted on an ecogram for south-west Ecuador (Figure 36). Only the species which have been recorded at least five times in south-west Ecuador were included; this was the case for only 11 species. Some species show rather narrow ecological amplitudes which fit well into the proposed vegetation classification (*T. multiflora, T. straminea, T. caerulea,* and *T. purpurea*) while others, notably *T. disticha, T. usneoides, T. floribunda* and *T. complanata,* have wide ecological amplitudes. Considering that only the most commonly recorded species and therefore those with the widest distribution and ecological ranges were included in this analysis, there seems to be a reasonably good correlation between the proposed classification and the distribution of bromeliads. A more detailed analysis of the little-recorded species would probably show that these have narrower ecological amplitudes and show a closer correlation. For example, the following additional Tillandsia species have only been recorded from Ceiba forests in the lowlands and foothills of south-western Ecuador and north-western Peru: *T. humilis, T. lehmanni, T. macrodactylon, T. pugiformis, T. schunkei* and *T. straminea.* They are probably 'typical' Tumbesian endemics.

Figure 34. Distribution of the 12 *Tillandsia* (Bromeliaceae) species which are endemic to south-west Ecuador. *T. aequatorialis. T. demissa. T. flagellata. T. fosteri. T. hauggii. T. marnieri-lapostollei. T. nervisepala. T. pseudotetrantha. T. rubro-violacea. T. rupicola. T. umbellata. T. zarumensis.* Three localities of *T. umbellata*, all within the study area, could not be located. See text for further comments.

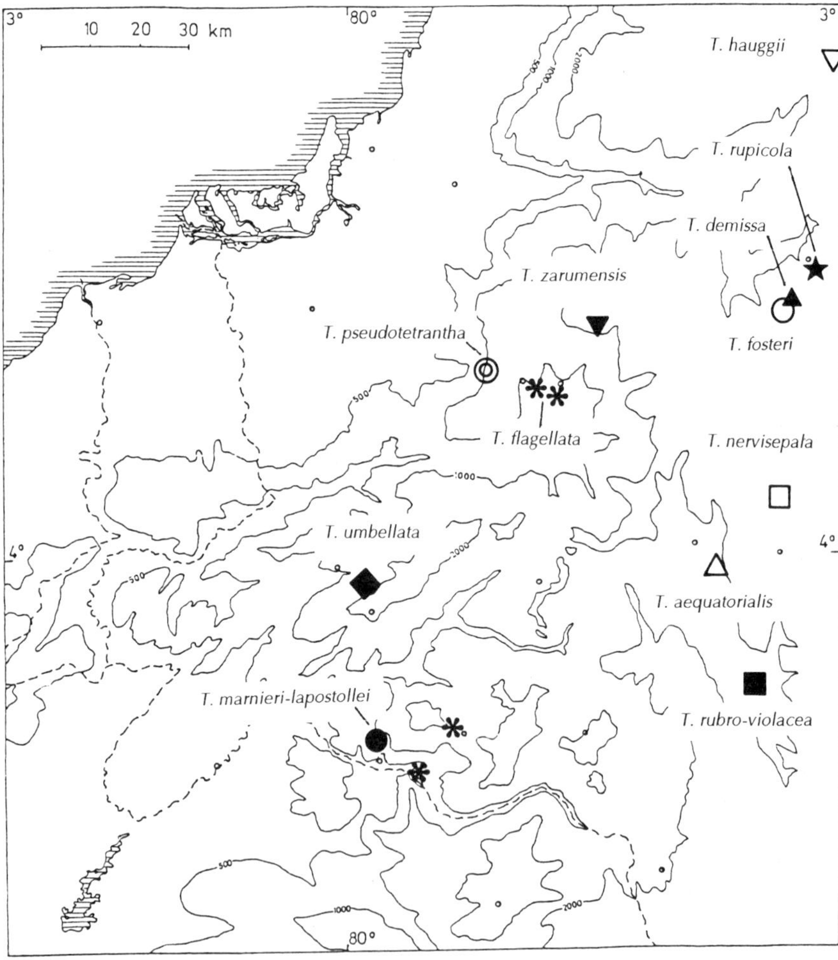

Figure 35. Ecological distribuion of 12 *Tillandsia* species which are endemic to south-west Ecuador. See Figure 32 for an explanation of the ecogram and text for further comments. Signatures as in Figure 34.

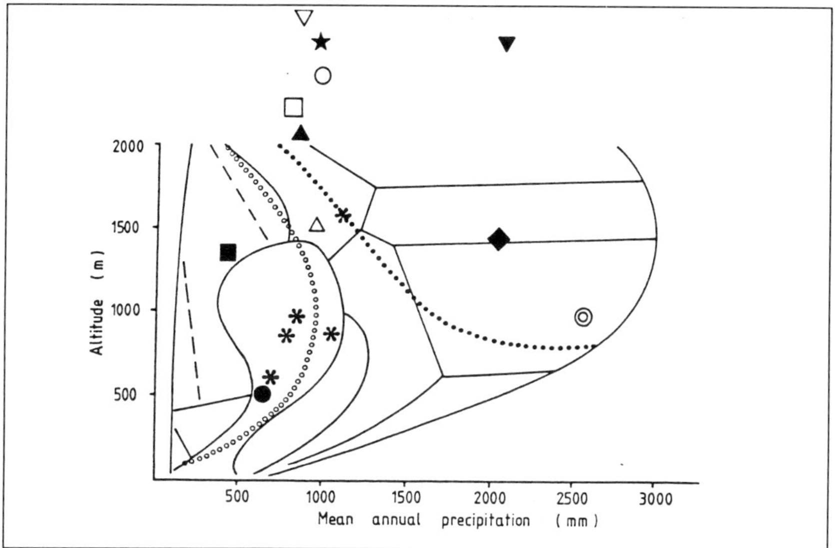

Conclusions

The distributional analyses of the Melastomataceae and Bromeliaceae confirm the general trends which were found in the analysis of the south-west Ecuadorian flora:

- It is a very species rich area, containing about 30% of the Ecuadorian flora.
- A considerable proportion (about 20%) of the plant species of the region are endemic to it.
- Local endemism is most pronounced at higher elevations, while lowland species tend to be distributed throughout the range of their respective vegetation types (Tumbesian endemics).

While it is too early to draw definitive conclusions, these data indicate that the provinces of Loja, El Oro and Azuay may in fact be exceptionally species-rich and certainly deserve particular attention in future studies.

These results compare well with those of Balslev (1988) who found that the mid-elevations between 900 and 3,000 m contain roughly half of the Ecuadorian flora and that about 40% of these are endemic to Ecuador. Furthermore, endemism

Figure 36. Ecological distribution of the 11 *Tillandsia* species which have been recorded five or more times in south-west Ecuador.
See Figure 32 for an explanation of the ecogram and the text for further comments.

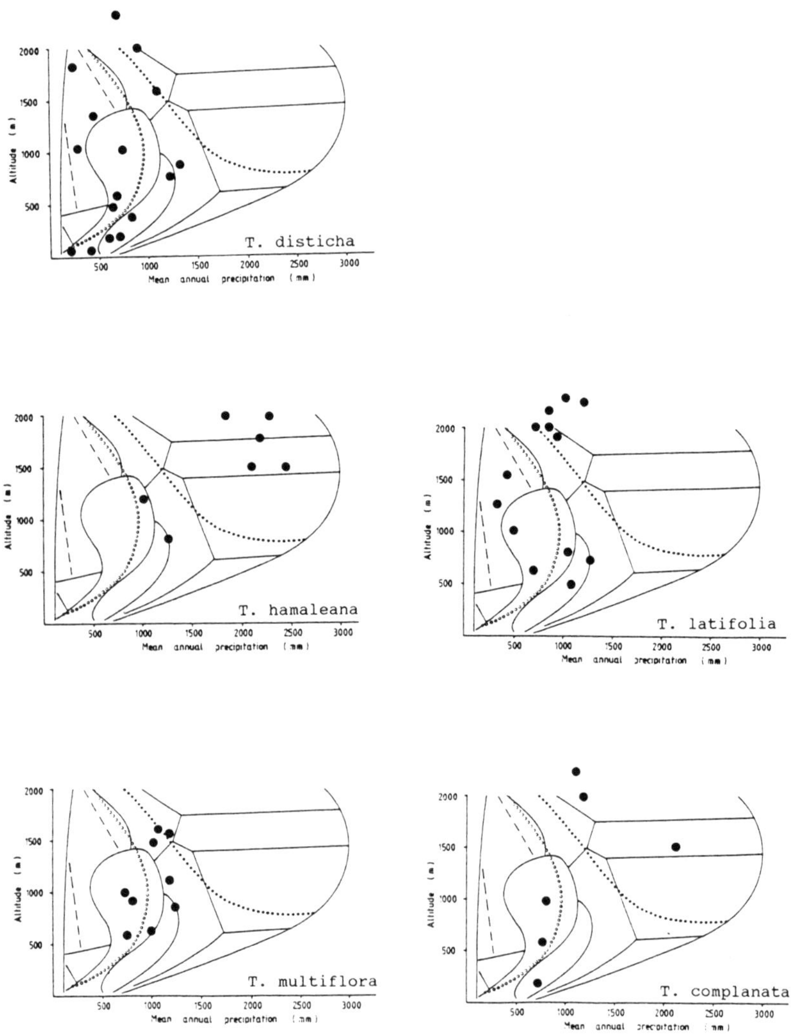

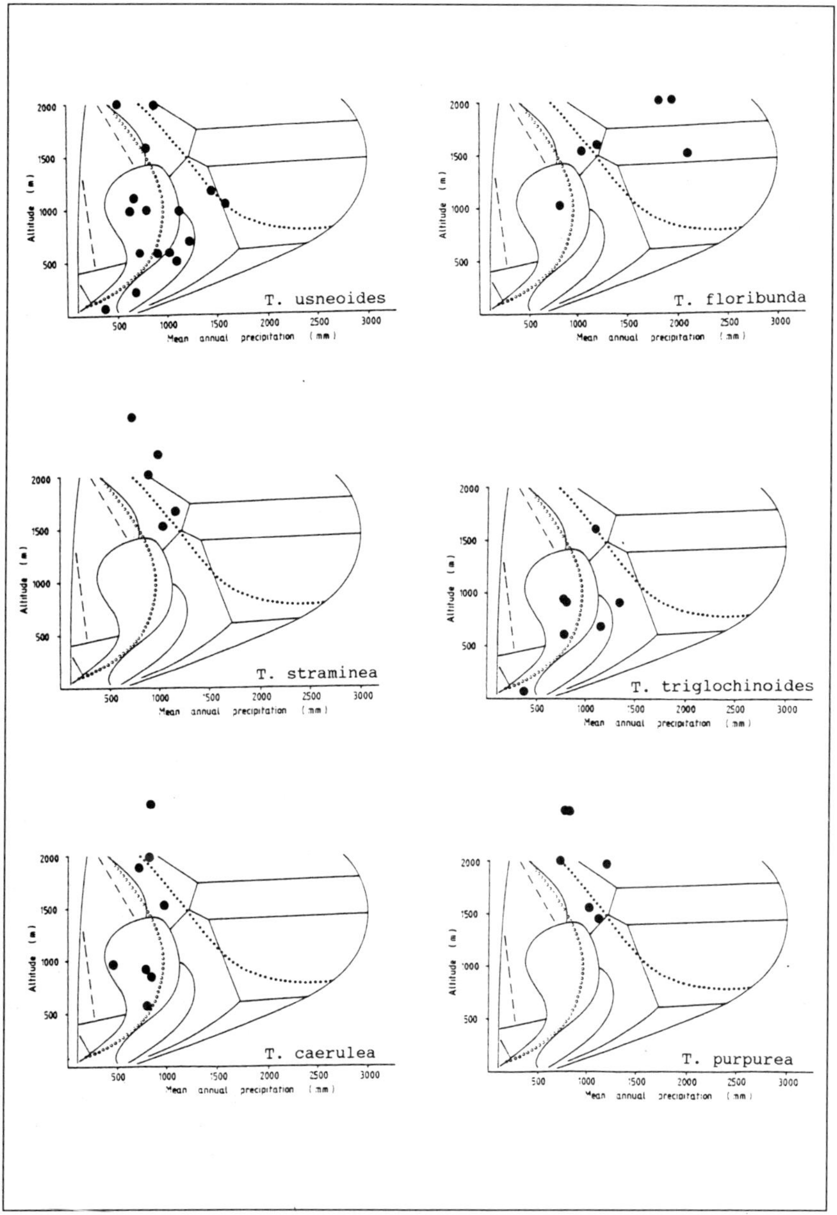
Altitude (m)
2000
1500
1000
500
500 1000 1500 2000 2500 3000
Mean annual precipitation (mm)
T. usneoides
T. floribunda
T. straminea
T. triglochinoides
T. caerulea
T. purpurea

was found to be much more pronounced west of the Andes than on the eastern side; this was attributed to the more varied climatic regimes ranging from deserts to pluvial forests as compared to the rather uniform Amazonian rainforests (Balslev 1988). This variety of habitats, which is especially conspicuous in south-west Ecuador, is further documented by the fact that the species overlap between the floras of Jauneche (Dodson et al. 1985) and Río Palenque (Dodson and Gentry 1978), barely 100 km further north, is only 27% (Balslev 1988). Dodson and Gentry (1991) estimated the proportion of species endemic to lowland western Ecuador (below 900 m) and adjacent areas of Colombia and Peru to be around 20%; this compares well with the 20% of endemics known from Capeira (Dodson and Gentry in prep.).

Unfortunately botanical studies are too few in the Tumbesian region to be able to test whether some areas, even within the same vegetation type, are richer in endemic species than others. Probably a considerable proportion of the dry forest plant species are distributed through a large proportion of the Tumbesian region, while the higher levels of local endemicity and diversity are concentrated on the Andean foothills and the Coastal Cordillera of Ecuador. Several new plant species and even a new tree genus have recently been discovered in the latter area (Parker and Carr 1992), emphasizing how poorly known it is.

HUMAN IMPACT

Human impact on the vegetation of the Tumbesian region is very marked, but tends to vary from one forest type to the next. Figure 37 shows the differing human uses of the vegetation types and their relative impact on them.

In arid areas grazing is the main threat and has a general impact on the whole ecosystem, for, when carried out within the forest it can completely suppress tree regeneration, leaving a forest which is a "living dead" (Janzen 1986). Agriculture is often limited to irrigable land in valley bottoms. Areas with 1,000 to 2,000 mm precipitation and a dry season of up to four months present the best conditions for agriculture and often are completely deforested, even on steep slopes. More humid regions tend to be less amenable for agriculture. High rainfall causes serious erosion, leaches the soil and provides optimal conditions for crop pests. Large tracts of land are therefore converted to cattle-pastures. The more obvious destruction of rainforests is one of the reasons why conservation efforts have focused on these rather than on tropical dry forests, whose destruction is equally severe, but more insidious (Janzen 1988). Indeed the tropical dry forests have been described as the most endangered of all the major lowland tropical forest habitats (Janzen 1988).

The remaining forest cover of south-west Ecuador was mapped based

Figure 37. Human influence on the different vegetation types in south-west Ecuador. Shading indicates the relative human population densities in the different zones. See text for further comments.

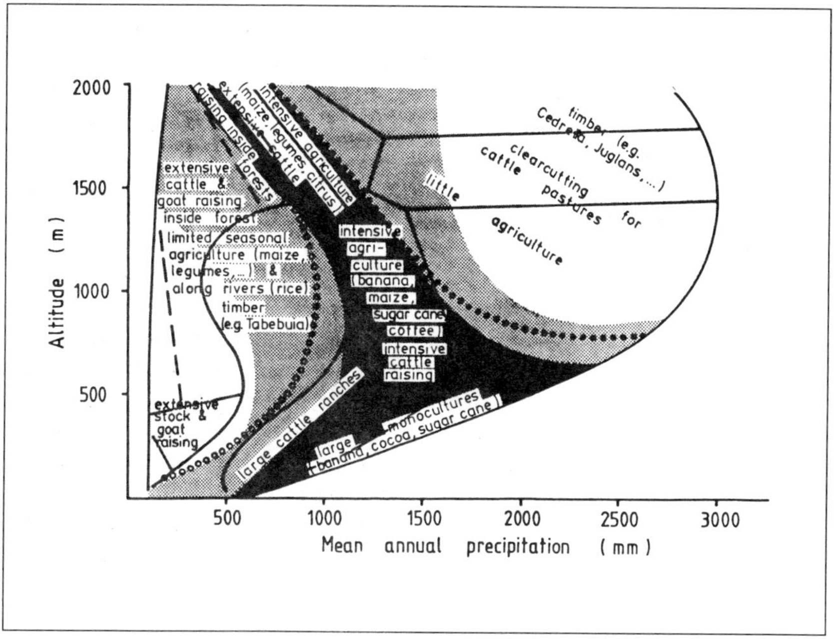

on observations made while travelling (Figure 14) and from LANDSAT images (2 Nov. 1986 and 26 Mar. 1987)[Figure 38]. While it is relatively easy to estimate forest cover in humid regions, it can be very difficult in arid areas, because large forest tracts may be present but heavily degraded and because dry forests are difficult to distinguish from scrubland on satellite images due to their lighter, less contrasting tone. The map therefore should be interpreted with caution.

Humid forest patches were mapped as of 1986 and 1987, but several of these were visited in 1991 and were found to be already largely fragmented. It has therefore to be feared that no natural forest larger than $10 km^2$ remains in the provinces of El Oro and Loja (with the exception of the Podocarpus National Park). Table 5 shows the estimated potential area occupied by each vegetation type and the remaining area of 'natural' forests. Forests were considered 'natural' if they are only slightly disturbed (e.g. by selective logging) and retain most of their original structure and species composition.

It is almost impossible to give respective estimates for the Tumbesian region as a whole, but as was already described above the situation is not much better. The only areas with substantial forest tracts remaining are the eastern half of the department of Tumbes in Peru and parts of the Coastal Cordillera in Guayas and Manabí Province, Ecuador. All lowland areas, especially those with a more humid climate are densely settled and completely deforested.

It is difficult to estimate the value forests have. Besides their importance to prevent erosion and for the protection of water-catchments, the plants themselves have multiple uses. Valverde (1991) gives an account of the uses given to the plants of the Chongón-Colonche Cordillera by the local inhabitants. Most important is the extraction of wood for construction, conversion to charcoal or use as firewood. Further forest products are tannins, resins and rubber. Other species (e.g. *Senna occidentalis, Cavanillesia platanifolia, Annona* sp., *Anacardium occidentale, Spondias* spp., *Inga* spp. and *Chrysophyllum cainito*) provide edible seeds and fruits, while an additional 55 species are listed as having medicinal properties.

The destruction of the mangrove forests, while not the main issue of this study, also deserves attention. Through the extensive construction of *camaroneras* (shrimp farms) starting in the late 1960s, by 1984 already over 10% of the mangrove area had been destroyed; in the Machala area alone 30% of the mangrove forests were lost between 1966 and 1982 (Jordan 1988). By the mid-1970s this had already lead to a scarcity of young shrimps, which are caught in the wild (Cintrón 1984). The remaining mangrove forests must be managed very carefully, if only for the survival of the shrimp industry which in 1986 accounted for about 10% of Ecuadorian exports, making it Ecuador's fourth most important export after petroleum, bananas and coffee (Jordan 1988).

Figure 38. Remaining forest cover in south-west Ecuador in 1987.
Determined by analysis of satellite images and field reconnaissance. Those forest patches arrowed were visited in 1991 and found to be degraded (Best 1992).

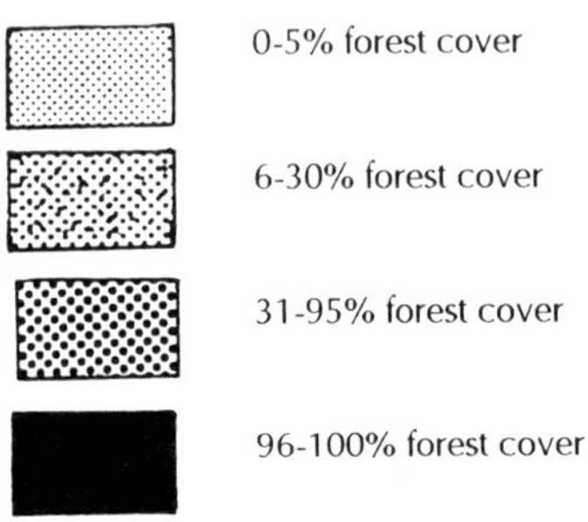

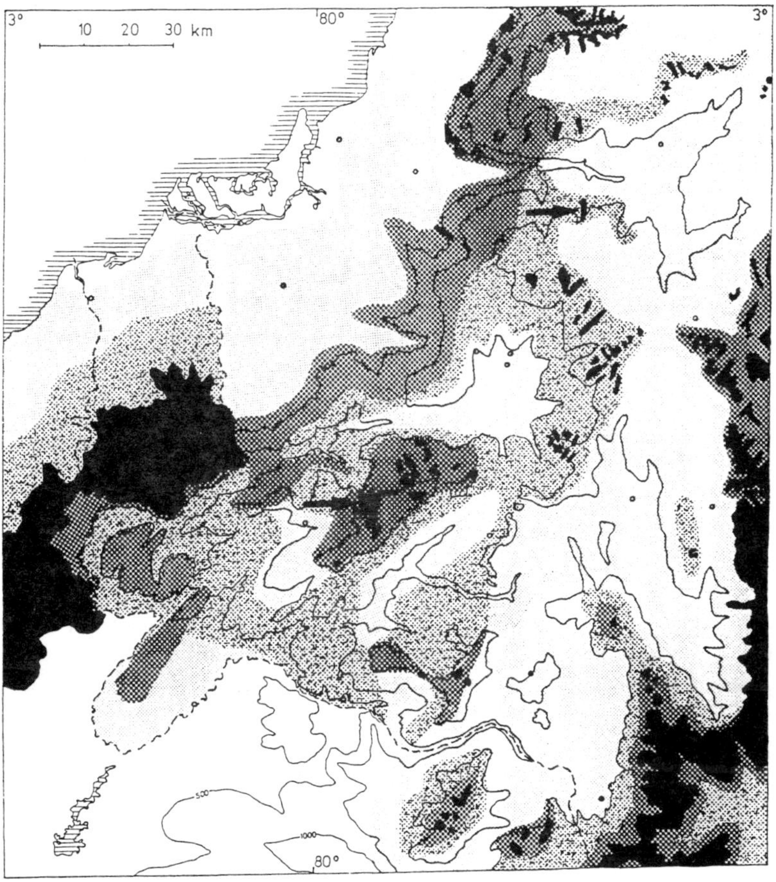
3°
10 20 30 km
80°
3°
500
1000
80°

CONSERVATION

The main problem throughout the Tumbesian region is how little forest there is left, particularly in the provinces of Loja, El Oro and Azuay (Figure 38, Table 5) and that there are few forest areas large enough to justify establishing a reserve. Generally it can be said that every single remaining forest patch plays a vital ecological role and also acts as a refuge for many widespread and a few threatened species and is therefore worth being preserved.

A possibility for the protection of these relict forest patches is the creation of Protection Forests set aside for the protection of watersheds, the prevention of erosion or as a future source of firewood. Such managed forests and 'buffer zones' which could be established on marginal land outside the reserves proper, can help to take pressure off the reserves.

Conservation priorities can be based on individual threatened species or on more general phytogeographic classifications. Unfortunately, far too little is known about the plants of the region to be able to assess the status of every single species. A list of the threatened plant species for Ecuador (IUCN 1990) even includes the important tree species *Ceiba trichistandra* and *Jacaranda sparrei* as threatened, but observations made in 1991 indicate that at least locally they show sufficient regeneration and that they are not in immediate danger of becoming extinct. For the Chongón-Colonche Cordillera, Valverde (1991) lists 36 species as threatened and states that 11 epiphytic plant species are being exterminated through the destruction of their host trees, while Kessler (in prep.) argues that only those bromeliad species which are restricted to rocky outcrops may be directly threatened through commercial collecting. However, this information is little more than anecdotal and much more detailed information is necessary on these, and the other 5,000+ plant species of the Tumbesian region. Also, while many individual plant species may well survive in disturbed or degraded forests, this is not equivalent to the protection of a habitat type or a whole ecosystem. Even the largest continuous forest area in the Tumbesian region, the North-West Peru Biosphere Reserve (226,300 ha), may be too small to support all plant and animal species on a long-term basis. For example, several bird species probably leave the reserve seasonally and move to other parts of the Tumbesian region. The ecological role of these species is unknown, and if they become extinct, some plant species which depend on the birds as pollinators or dispersers, might follow soon.

For these reasons, and since we do not have the time to wait until more information is available, it seems prudent to base the priorities for conservation on a phytogeographical classification such as the one proposed here. Apart from forest area remaining in south-west Ecuador other criteria for setting conservation priorities are global distribution (uniqueness) and degree of representativeness of particular communities, number of species within a community and degree of endemism. According to Terborgh and Winter (1983) and Fjeldså (1991)

Table 5. Estimated area covered by each vegetation type in Loja, El Oro and Azuay (north to 3°S) Provinces, Ecuador below 2,000 m (based on Figure 29) and approximate percentage cover of each (based on Figure 38). Note that the more arid vegetation types tend to be more heavily deforested than the mesic ones.

Vegetation type	Area km^2	Percentage area with forest cover 0-5%	6-30%	31-95%	96-100%
Mainly Deciduous Tropical Thorn-forest and *Acacia* Thorn-Forest	650	85	10	5	-
Mainly Deciduous *Ceiba trichistandra* Forest	3,000	20	70	10	-
Semi-evergreen *Ceiba pentandra* Forest	1,600	80	15	5	-
Semi-evergreen Lowland and Premontane Tall Forest	1,700	50	25	24.5	0.5
Moist Lowland Forest	850	75	-	23	2
Humid to Very Humid Premontane Cloud-forest	1,000	-	3	94	3
Deciduous to Semi-evergreen Intermontane Scrub, Thorn-forest and Forest	3,500	100	-	-	-
Humid to Very Humid Lower Montane Cloud-forest	700	5	50	42	3
Deciduous to Semi-evergreen Lower Montane Cloud-forest	300	-	35	65	-
Humid to Very Humid Montane Cloud-forest (below 2,000 m)	500	-	65	33	2
Total	**13,800**	**54**	**26.5**	**19**	**0.5**

endemicity should take priority over species number as a criterion for designating areas of conservation priority.

As was found in the analysis of the phytogeography of the region, relatively few species occurring below 2,000 m have restricted ranges within the distribution of their respective vegetation types. Higher up in the mountains topographical barriers attain greater importance, often leading to local endemism on mountain ranges. Thus, to preserve the total flora of the region, the ideal strategy would be to protect representative samples of each lowland vegetation type and several examples of montane cloud-forest and páramo vegetation. For the latter the most interesting areas would probably be:

- The mountains east and south-east of Loja (fortunately already protected as the Podocarpus National Park) (Bloch *et al.* 1991).
- The Cajas area in Azuay (part of which is set apart as the Cajas National Recreation area).
- The Chilla mountains.
- The Cordillera Cordoncillo east of Saraguro.

Conservation priorities

Based on the criteria of uniqueness, endemicity and species number, the thirteen vegetation types found in the Tumbesian region have been assigned to three priority classes for conservation. This section attempts to provide a basis for setting conservation priorities; the identification of particular key areas and a discussion of the logistic and socio-economic problems associated with conservation projects in the area will be made in the conservation recommendations chapter.

GROUP 1
Mainly Deciduous Tropical Thorn-forest and *Acacia* Thorn-forest
Mainly Deciduous *Ceiba trichistandra* Forest
Semi-evergreen *Ceiba pentandra* Forest
Semi-evergreen Lowland and Premontane Tall Forest

Together these four vegetation types constitute the central components of the 'Tumbesian Dry Forest', and their preservation has to be the main concern in any conservation plan for the Tumbesian region. The proportion of endemic plant species was estimated by Dodson and Gentry (1991) to be about 20%.

GROUP 2
Humid to Very Humid Premontane Cloud-forest
Deciduous to Semi-evergreen Lower Montane Cloud-Forest
Humid Coastal-hill Cloud-forest
Deciduous to Semi-evergreen Intermontane Scrub, Thorn-forest and Forest

These four forest types have very limited global distributions along the edges of the Tumbesian centre. The establishment of conservation areas within the provinces of Loja, El Oro, Azuay, Guayas and Manabí (all in Ecuador) is of vital importance to safeguard them.

Humid to Very Humid Premontane Cloud-forest is certainly the vegetation type with the smallest overall range as it is restricted to a narrow altitudinal belt (500-1,500 m, often less) along a 125-km stretch of the western Andean slope. At least two endemic bird species are known to be restricted to this vegetation type and there is certainly a large number of endemic plant species. It also has the highest species diversity.

Deciduous to Semi-evergreen Lower Montane Cloud-Forest is another special vegetation type with a very limited distribution, both in the study area and world-wide (Mueller-Dombois and Ellenberg 1974). A fairly high level of endemism and the unique combination of (semi)-deciduousness and abundant epiphytic growth clearly set this forest type apart.

Humid Coastal-hill Cloud-forest occurs only on the coastal hills of Manabí and Guayas, and probably supports a number of endemic species. Ongoing studies in the Machalilla National Park (C. Josse, verbally 1992) will hopefully soon provide a database on which to judge the importance of this vegetation type.

Deciduous to Semi-evergreen Intermontane Scrub, Thorn-forest and Forest is a conglomerate of vegetation types that has been so adversely affected by human activities that probably only one patch of about 30 ha remains at Hacienda Quesada in the upper Rircay valley. It might support some local species and its immediate preservation (even before the forest patch is studied in detail) might be essential for the survival of these species.

GROUP 3
Desert
Semi-desert
Moist Lowland Forest
Humid to Very Humid Lower Montane Cloud-forest
Humid to Very Humid Montane Cloud-forest

These five vegetation types reach the Tumbesian region only peripherally, the latter occurring mainly above 2,000 m. They do not form central components of the Tumbesian region and while some Tumbesian species certainly also occur in them, they probably do not have high numbers of Tumbesian endemics. The conservation of these vegetation types is certainly better achieved elsewhere.

■

AVIFAUNA

PREVIOUS ORNITHOLOGICAL WORK IN THE TUMBESIAN REGION

ALTHOUGH A few early bird collectors visited the Tumbesian region in the late 19th century (e.g. Taczanowski from 1884 to 1886), it was Frank Chapman and his colleagues from the American Museum of Natural History who first studied its distinctive birdlife in detail, during the first two decades of the 20th century. In his monumental synthesis on the avifauna of Ecuador, he recognized that the area supported a large number of species whose ranges fell entirely within western Ecuador and adjacent north-western Peru (Chapman 1926). He named these species the "Equatorial Arid Fauna", as many of them appeared to be adapted to the arid scrub and desert-like conditions of coastal Ecuador and Peru.

From 1930 to 1970 very few ornithologists visited the Tumbesian region, the most notable work being by Marchant (1958), Koepcke (1961) and D. Norton and R. A. Paynter in 1964 and 1965. It was only in the late 1970s and 1980s that surveys recommenced, and papers concerning the avifauna of the area began to reappear (Schulenberg and Parker 1981, Parker *et al.* 1985, Wiedenfeld *et al.* 1985, Parker *et al.* 1989, Robbins and Ridgely 1990). When a completely new species was discovered in south-west Ecuador in the early 1980s (*Pyrrhura orcesi*: Ridgely and Robbins 1988) it became clear that there were still large gaps in the ornithological knowledge of the area. Since the late 1980s there has been an up-surge in ornithological interest in the Tumbesian region and several individuals and institutions have undertaken field research there (Best and Clarke 1991, Bloch *et al.* 1991, Krabbe 1991, Best 1992, Parker and Carr 1992, Williams and Tobias 1994; also unpublished work by the Academy of Natural Sciences of Philadelphia and the Western Foundation of Vertebrate Zoology).

THE RESTRICTED-RANGE BIRDS OF THE TUMBESIAN CENTRE OF ENDEMISM

Complementary to the fieldwork described above have been 'desk-studies' on the avian biogeography of the area, notably by Cracraft (1985), who identified 37 species whose concurrent and restricted ranges constituted an area of endemism which he named the "Tumbesan Centre" after its geographical centre, the Department of Tumbes in north-west Peru. In this book we use Ridgely and Tudor's (1989) revised name, the "Tumbesian centre". Müller (1973) studied the terrestrial vertebrates (including the birds) of the same region which he distinguished as a faunal centre and gave the name the " Ecuadorian Subcentre" ; this formed part his "Andean Pacific Centre" .

The most recent investigations into the restricted-range bird species of the area have been conducted by BirdLife International's Biodiversity Project, as part of a global classification of "Endemic Bird Areas" (ICBP 1992, Stattersfield *et al.* in prep.). The Tumbesian Western Ecuador and Peru EBA has one of the highest numbers of endemic bird species of any South American EBA. Fifty-five species with ranges smaller than 50,000 km^2 occur in the region, with 46 of these entirely confined to it (Table 6). The area stood out as "Critical" in BirdLife International's priority ranking of EBAs on a rising scale of "High", "Urgent" and "Critical"(ICBP 1992).

Western South America has an unusually high density of EBAs and the Tumbesian region is important not only as a rich centre of avian endemism in its own right, but it is also significant because it meets with two other areas of bird endemism, which combine to bring a highly distinctive mix to the avifauna of the region, composed of a large proportion of restricted-range species. Figure 39 shows the position of the Tumbesian EBA with respect to its two nearest neighbours: the Chocó and Pacific slope Andes EBA which overlaps with the northern part of the Tumbesian EBA; and at higher elevation, the South Central Andean forests EBA which overlaps to the south. The altitudinal limits and habitat types of these EBAs are shown in Table 6.

Table 6. Altitudinal range and habitat types of three Neotropical Endemic Bird Areas.

EBA	Number of restricted range species	Altitudinal range	Habitats
Chocó and Pacific slope Andean	62	sea-level to 3,000 m	wet forest
Tumbesian western Ecuador and Peru	55	sea-level to 3,000 m (mostly below 2,000 m)	wet and dry forest
South Central Andean Forest	8	1,500-3,500m	cloud-forest

Source: Stattersfield *et al.* (in prep.).

Figure 39. Three Endemic Bird Areas in Ecuador and Peru.
Adapted from ICBP (1992) and Stattersfield *et al.* (in prep.).

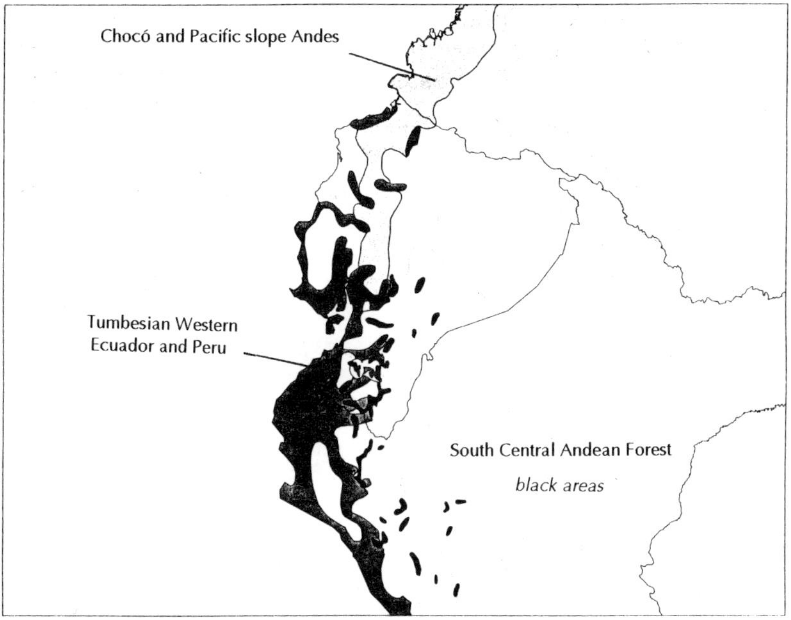

Figure 39 gives a general outline of the area covered by each EBA; it should be noted that some restricted-range birds go beyond these boundaries, into neighbouring EBAs. As would be expected, most overlap occurs closest to the zone of contact between two EBAs, but habitat and altitude have important roles to play in determining which species from other EBAs occur at particular sites in the Tumbesian region.

The effects of habitat

Habitat type strongly influences the distribution of restricted-range species in the Tumbesian region. The Chocó and Pacific slope Andes EBA reaches its southern limit in south-western Ecuador, and species from that EBA occur here only in very humid forests, which tend to grow only in a narrow altitudinal band where the climate is suitable (page 69). This results in a rather localized distribution of these species within south-western Ecuador at such sites as Buenaventura in El Oro Province, which possesses forest sufficiently evergreen to support 7 Chocó endemics (Table 7). Tumbesian endemics, by contrast, tend to avoid the most humid forest sites such as Buenaventura, favouring deciduous to semi-evergreen sites such as Tambo Negro in Loja Province (Table 7). Here there are more than twice as many Tumbesian endemics than at Buenaventura, which supports only those which occur in the more humid vegetation types of the region, together with a few arid Tumbesian species in degraded parts of the site.

Such overlap zones have special conservation significance because they can support unusually large numbers of restricted-range species, and well-placed reserves could protect species from more than one centre of endemism (Terborgh and Winter 1983).

Table 7. Numbers of restricted-range bird species from three EBAs found at four sites in south-west Ecuador.

Locality/coordinates	Altitude	Habitat	Tumbesian	Chocó/ Pacific slope	S. Central Andean
Tambo Negro, Loja Prov. 4°24'S 79°51'W	500-1,000 m	Deciduous *Ceiba trichistandra* Forest	26	0	0
Sozoranga, Loja Prov. 4°21'S 79°45'W	1,300-2,000 m	Semi-evergreen Lower Montane Cloud-forest	21	0	2
Buenaventura, El Oro Prov. 3°40'S 79°44'W	900-1,000 m	Very Humid Premontane Cloud-forest	9	7	1
Utuana, Loja Prov. 4°22'S 79°43'W	2,500 m	Humid Montane Cloud-forest	7	0	2

Sources: Robbins and Ridgely (1990), Best and Clarke (1991), Krabbe (1991), Best *et al.* (1992, 1993).

Species from the South Central Andean forests EBA occur mainly along the south-eastern edge of the Tumbesian region, where there is humid forest in the altitudinal range 1,500-3,200 m. These species also occur more centrally in the Tumbesian region, where they can descend as low as 1,000 m or less where the forest is sufficiently humid (e.g. at Buenaventura).

The effects of altitude

The strong influence of altitude can be seen by comparing the avifauna of two localities in Loja Province: Sozoranga and Utuana, based on two recent surveys (Best and Clarke 1991, Best *et al.* 1992; Table 7). Sozoranga is situated between 1,300 and 2,000 m, whereas Utuana lies somewhat higher at 2,500 m, beyond the upper altitudinal limit of most of the Tumbesian endemics. This difference in altitude may have accounted for a 72% drop in Tumbesian endemics found at Utuana despite the fact that the two localities lie only 7 km apart.

THE CONSERVATION STATUS OF THE TUMBESIAN AVIFAUNA

The recent up-surge in ornithological fieldwork in the Tumbesian region has greatly improved our knowledge of its avifauna and focused attention on its threats. A much clearer picture has emerged of which species are in danger, culminating in their classification in the recently published *Threatened birds of the Americas* (Collar *et al.* 1992). This chapter discusses the status and conservation of the endemic avifauna of the Tumbesian region.

The special case of birds with restricted ranges

A fundamental principle when assessing the conservation status of the Tumbesian avifauna (and that of other EBAs) is that, because the ranges of its component species are geographically small, they are usually more vulnerable than are their more wide-ranging counterparts. Although population density is important, they are often more vulnerable to local habitat change and they tend to have smaller overall populations to replenish areas where population declines have occurred. In the Tumbesian region several species have particularly small ranges (e.g. a few hundred km^2 in the case of *Penelope albipennis*) even compared to the rest of the Tumbesian endemics. The vulnerability of such species is often inversely proportional to the size of their ranges.

The habitat preferences of the Tumbesian avifauna

The 'Vegetation' chapter showed that whereas some of the Tumbesian habitats are seriously threatened, others are not at risk. Similarly some Tumbesian endemics are secure. Those species which are reliant on the habitat types which have been or are being severely degraded are most at risk. Those of scrub or degraded forest are relatively 'safe' at present, unless some other factor (such as the bird trade) seriously affects their populations. The area of such degraded habitats in the Tumbesian region is increasing. Although detailed habitat and altitudinal preferences of most restricted-range bird species of the Tumbesian region are not yet established due to lack of data, the broad requirements of each species are known (Table 8). The list includes the 46 restricted-range species confined to the region, together with the nine species which are shared between the Tumbesian EBA and other EBAs. The table also lists the nature and severity of the threats which they face.

The number of Tumbesian endemics recorded from each of six habitat types is shown in Figure 40. The species are broadly distributed across the habitat types, with the deciduous forest and scrub categories providing habitats for the highest numbers of Tumbesian endemics (38 species [68%] and 37 species [66%] respectively), closely followed by the semi-evergreen forest (28 species; 50%),

Table 8. The avifauna of the Tumbesian Centre of Endemism.

English name	Scientific name	Altitude (m)	Habitats	THREATS				
				deforestation	understorey disturbance	hunting	trade	tiny range
Pale-browed Tinamou	*Crypturellus transfasciatus*	0-1,000	D, S, (E, C)	X	-	X	-	-
Grey-backed Hawk	*Leucopternis occidentalis*	0-2,900	S, F	XX	-	-	-	-
Rufous-headed Chachalaca	*Ortalis erythroptera*	0-1,850	D, S, E	X	-	X	-	-
White-winged Guan	*Penelope albipennis*	0-800	D, (M)	X	X	X	-	XX
Ecuadorian Ground-Dove	*Columbina buckleyi*	0-1,000	D, G	-	-	-	-	-
Ochre-bellied Dove	*Leptotila ochraceiventris*	0-2,625	D, S, F, (G)	XX	XX	X	-	-
Red-masked Parakeet	*Aratinga erythrogenys*	0-2,500	D, S, E, G, A	X	-	-	X	-
El Oro Parakeet	*Pyrrhura orcesi*	300-1,300	E	XX	-	-	-	X
Pacific Parrotlet	*Forpus coelestis*	0-2,150	D, S, C, A	-	-	-	-	-
Grey-cheeked Parakeet	*Brotogeris pyrrhopterus*	0-1,400	D, S, A (F)	X	-	-	X	-
Scrub Nightjar	*Caprimulgus anthonyi*	0-750	D, G, C	-	-	-	-	-
Tumbes Hummingbird	*Leucippus baeri*	0-1,300	D, C	-	-	-	-	-
Short-tailed Woodstar	*Myrmia micrura*	0-1,000	D, C	-	-	-	-	-
Esmeraldas Woodstar	*Acestrura berlepschi*	0-150	S, G	X	X	-	-	X
Ecuadorian Piculet (2)	*Picumnus sclateri*	0-2,000	D, S, C	-	-	-	-	-
Surf Cinclodes+	*Cinclodes taczanowskii*	0-100	S, O	-	-	-	-	-
Coastal Miner+	*Geositta peruviana*	0-300	S, O	-	-	-	-	-
Necklaced Spinetail+	*Synallaxis stictothorax*	0-560	D, S	-	-	-	-	-
Blackish-headed Spinetail	*Synallaxis tithys*	0-1,000	D, S, (E), (C)	XX	XX	-	-	-
Rufous-necked Foliage-gleaner (2)	*Syndactyla ruficollis*	400-2,900	D, S, E, G	X	XX	-	-	-
Henna-hooded Foliage-gleaner (2)	*Hylocryptus erythrocephalus*	200-1,750	D, S, G, (E)	XX	XX	-	-	-
Collared Antshrike+ (3)	*Sakesphorus bernadi*	0-2,000	D, S, C	-	-	-	-	-
Chapman's Antshrike	*Thamnophilus zarumae*	400-2,625	G, C, (D, S, E)	-	-	-	-	-
Grey-headed Antbird	*Myrmeciza griseiceps*	600-2,900	S, E, (D)	X	XX	-	-	-
Watkins's Antpitta	*Grallaria watkinsi*	0-2,000	D, S, E, G, C	-	-	-	-	-
Elegant Crescent-chest	*Melanopareia elegans*	0-2,300	G, C	-	-	-	-	-
Grey-and-white Tyrannulet	*Pseudelania leucospodia*	0-600	C	-	-	-	-	-
Pacific Elaenia	*Myiopagis subplacens*	0-1,500	D, S, E, G	-	-	-	-	-

cont.

Table 8 (cont.)

Pacific Royal-flycatcher	*Onychorhynchus occidentalis*	0-900	D, S, E, G	XX	XX	-	-	-
Grey-breasted Flycatcher+	*Lathrotriccus griseipectus*	0-1,750	D, S, E	XX	XX	-	-	-
Piura Chat-tyrant	*Ochthoeca piurae*	1,500-2,800	C	-	-	-	-	-
Tumbes Tyrant	*Tumbezia salvini*	0-1,000	D, C	-	-	-	-	-
Ochraceous Attila+	*Attila torridus*	0-1,700	S, E	XX	XX	-	-	-
Rufous Flycatcher	*Myiarchus semirufus*	0-500	C	-	-	-	-	-
Sooty-crowned Flycatcher+ (2)	*Myiarchus phaeocephalus*	0-2,000	D, C	-	-	-	-	-
Baird's Flycatcher	*Myiodynastes bairdi*	0-500	D, C	-	-	-	-	-
Slaty Becard+	*Pachyramphus spodiurus*	0-825	D, S, G, (E)	XX	XX	-	-	-
Peruvian Plantcutter	*Phytotoma raimondii*	0-500	C	-	X?	-	-	XX
White-tailed Jay	*Cyanocorax mystacalis*	0-1,850	D, S, G, C	-	-	-	-	-
Superciliated Wren	*Thryothorus superciliaris*	0-500	D, C, A	-	-	-	-	-
Plumbeous-backed Thrush	*Turdus revei*	0-2,300	D, S, G, C	-	-	-	-	-
Ecuadorian Thrush	*Turdus maculirostris*	0-2,200	D, S, G, S	-	-	-	-	-
Three-banded Warbler	*Basileuterus trifasciatus*	900-3,050	S, E, G	-	-	-	-	-
Grey-and-gold Warbler (2)	*Basileuterus fraseri*	0-1,900	D, S, E, G	-	-	-	-	-
Black-cowled Saltator	*Saltator nigriceps*	1,000-2,900	S, E, G, C	-	-	-	-	-
White-headed Brush-Finch	*Atlapetes albiceps*	400-1,300	DF, C	-	-	-	-	-
Bay-crowned Brush-Finch (3)	*Atlapetes seebohmi*	1,200-2,600	S, E, G, C	-	-	-	-	-
Pale-headed Brush-Finch	*Atlapetes pallidiceps*	1,500-2,100	S, C, A	XX?	XX?	-	-	XX
Drab Seedeater+	*Sporophila simplex*	0-1,500	C, A	-	-	-	-	-
Sulphur-throated Finch	*Sicalis taczanowskii*	0-200	C, A	-	-	-	-	-
Cinereous Finch	*Piezorhina cinerea*	0-300	S	-	-	-	-	-
Crimson Finch-Tanager	*Rhodospingus cruentus*	0-750	D, C, A	-	-	-	-	-
Black-capped Sparrow+ (2)	*Arremon abeillei*	0-1,750	D, S, G, C	-	-	-	-	-
Tumbes Sparrow	*Aimophila stolzmanni*	0-1,950	D, C	-	-	-	-	-
White-edged Oriole	*Icterus graceannae*	0-1,500	D, G, C	-	-	-	-	-
Saffron Siskin	*Carduelis siemiradzkii*	0-750	D, C	XX?	X?	-	-	-

Numbers in parentheses after the species name refer to the number of subspecies present in the Tumbesian region. + occurs in another EBA. * species not listed in Stattersfield *et al.* (in prep.) but included here (see Table 3).
Habitats: D= deciduous forest, S= semi-evergreen forest, E= evergreen forest, G= secondary growth, C=scrub, A= agricultural land, M=mangroves, O= coastal habitats. Habitats in parenthesis of lesser importance. **Threats:** XX = major threat: a primary cause of a species being threatened, X=minor threat: a secondary threat to a threatened species, or a primary threat to a near-threatened species.

Sources: Ridgely and Tudor (1989), Collar *et al.* (1992), P. Coopmans *in litt.* (1992), C. Rahbek *in litt.* (1992), M. B. Robbins *in litt.* (1992), J. Tobias *in litt.* (1992), Ridgely and Greenfield (in prep.), Stattersfield *et al.* (in prep.) and personal observations.

secondary growth (24 species; 43%) and evergreen forest (22 species; 39%) cacategories.

The majority of Tumbesian species occur in both forest and non-forest habitat types (Figure 41), with nine (16%) being restricted to forest and 10 species (18%) confined to non-forest habitats. Table 8 shows that these non-forest species currently face no threats, so they can be considered 'safe' at present. Conversely, the nine which occur only in forest are the most threatened, and Table 8 shows that each faces at least one major threat.

Figure 42 shows the importance of forest to the 22 threatened and near-threatened Tumbesian species (subsequently called the "priority Tumbesian bird species"): most of these species occur in forest; seven species have also been found in scrub, seven in secondary growth and three in agricultural land. It is not known whether these species can breed in such habitats though.

Figure 40. Numbers of Tumbesian endemics recorded in six habitat types.
Note the broad range of habitat types occupied.

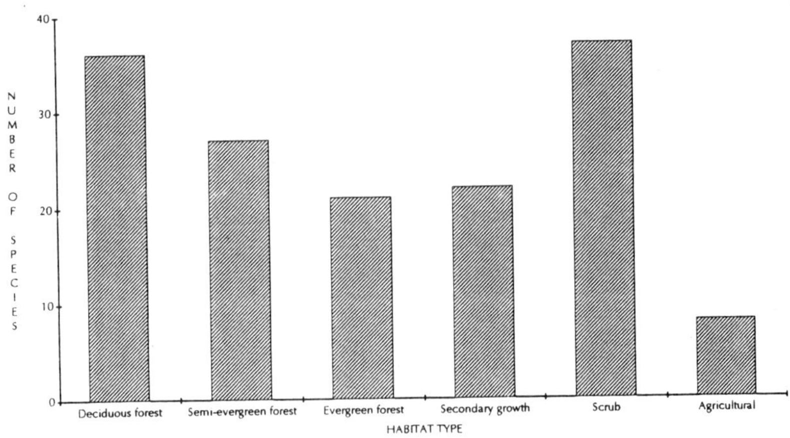

Source: Table 8.

Figure 41. Proportion of Tumbesian endemics which occupy forest only, scrub only and mixed habitats.

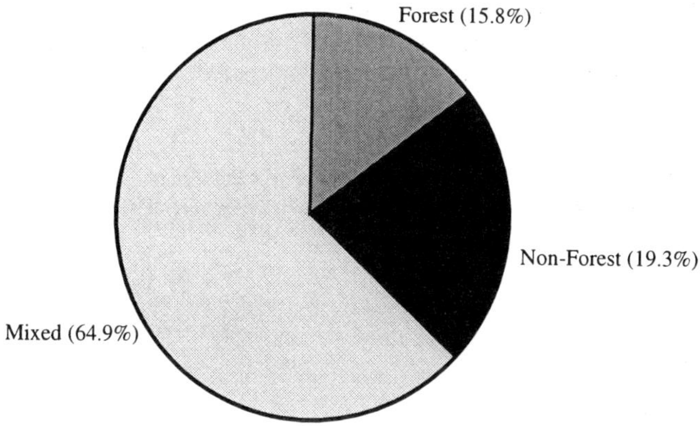

Source: Table 8.

Figure 42. Numbers of threatened and near-threatened Tumbesian endemics occurring in six habitat types. The importance of forests to these species is clear.

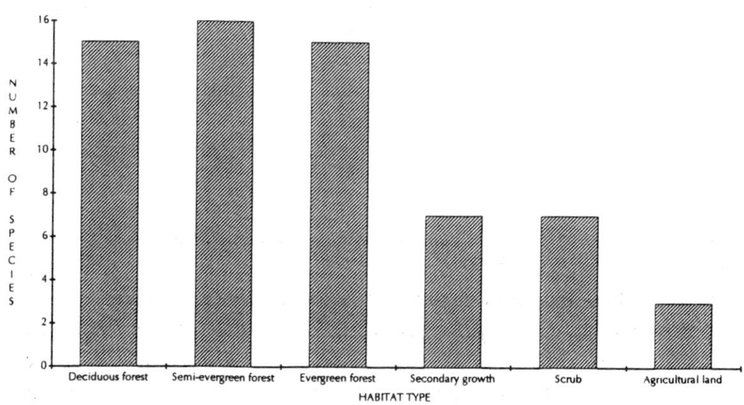

Source: Table 8.

THREATS TO THE TUMBESIAN AVIFAUNA

The Tumbesian avifauna is affected by five different threats (Table 8). Figure 43 shows the comparative importance of each of these. The most severely endangered Tumbesian endemics are those species which suffer a combination of threats such as *Penelope albipennis*. This species is imperiled by both habitat destruction and hunting, compounded by its tiny range and population, placing it in a critical position.

Figure 43. Number of Tumbesian endemics affected by five threats.
Deforestation and understorey degradation stand out as the dominant threats.

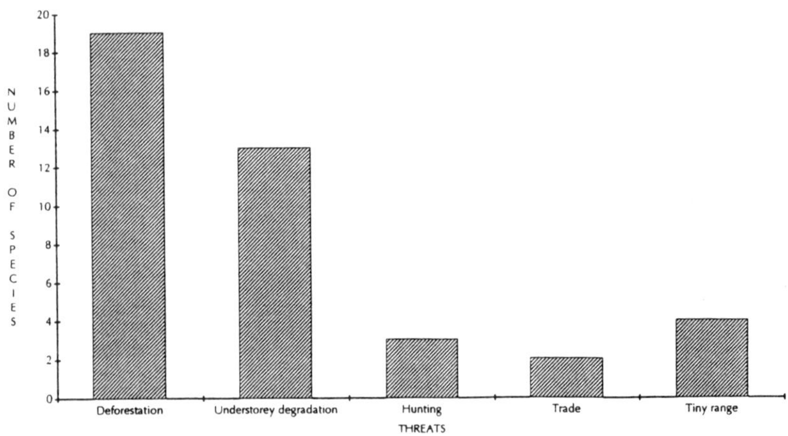

Source: Table 8.

Habitat destruction

Deforestation and understorey disturbance (both clearing and trampling) are by far the most serious threats to the Tumbesian avifauna, affecting 19 (86%) and 13 (59%) respectively of the priority species. They often occur simultaneously when a forest is cut down, but many Tumbesian forests retain large, intact trees above an understorey heavily degraded by grazing animals or disturbed by firewood gathering. This has important implications for several apparently understorey dependent endemics (e.g. *Leptotila ochraceiventris*, *Synallaxis tithys* and *Myrmeciza griseiceps*). Such forests show up as intact on satellite images and aerial photographs, but are unsuitable for understorey-dependent species.

The broad habitat categorizations of Table 8 have been used because a detailed vegetation classification is only available for part of the Tumbesian region. They hide the fact that it supports an unusually rich variety of habitat types, including no less than 10 distinct kinds of forest . Because some forest-reliant birds of the region occupy only a proportion of these forest types, they are vulnerable to even partial, localized deforestation. The habitat preferences of the threatened species are presented in diagrams later in this chapter, which are based on the vegetation classification already presented.

Hunting

Hunting affects three (14%) priority species. Although many bird species are hunted opportunistically in Ecuador and Peru, including ducks, hawks, parrots, trogons, toucans and even the smaller species which are killed by children with sling-shots and catapult (P. Greenfield *in litt.* 1992, M. B. Robbins *in litt.* 1992), the principal species taken are tinamids, cracids and columbids, which are killed for their meat. Their eggs are also sometimes collected by local people. *Crypturellus transfasciatus, Ortalis erythroptera, Penelope albipeñnis* and *Leptotila ochraceiventris* are the Tumbesian endemics most at risk from hunting; *Leucopternis occidentalis* may also be occasionally shot for sport rather than its meat value. In general mammals have been much preferred over birds.

The effects of hunting on the priority species are difficult to quantify as very few quantitative data exist. A man was seen with four dead *Ortalis erythroptera* near Atacames (Manabí Province) in coastal western Ecuador (F. Ortiz-Crespo *in litt.* 1991); the same species has been shot by border guards in the Tumbes National Forest, Peru, apparently because they had only rice to eat (M. B. Robbins *in litt.* 1992). A small basket-like trap for catching tinamous was found above Sozoranga in Loja Province, Ecuador (C. T. Clarke *in litt.* 1992). The relatively high cost of guns and ammunition probably makes the practice uneconomic for many local people. This would explain why the groups of *Ortalis erythroptera* frequently found calling loudly from forest patches along well-used roads seem not to attract hunters. However, the extent of hunting probably varies locally, and in some areas the species is known to call at night, thereby avoiding day-time hunting.

Cracids are also occasionally captured because locals mistakenly believe they can be cross-bred with domestic chickens to produce super-strong fighting cocks (R. S. R. Williams verbally 1991). *Ortalis erythroptera* is known to have been captured for this purpose as has the more montane Bearded Guan *Penelope barbata* from the South Central Andean forests EBA, the latter species at Amaluza (Loja Province, Ecuador) at the eastern edge of the Tumbesian region (Williams and Tobias 1994).

Parrot trade

Two of the four parrots endemic to Tumbesian region (*Aratinga erythrogenys* and *Brotogeris pyrrhopterus*) are traded locally and internationally. Both have been found in local homes (Best and Clarke 1991, Williams and Tobias 1994), and *Aratinga erythrogenys* is regularly sold (bleached or dyed yellow, perhaps to make them look more exotic) in the streets in Quito and Guayaquil (P. Greenfield *in litt.* 1992). The international trade is potentially a much graver threat; both have been in trade for many years (Inskipp and Corrigan 1992). The population declines noted in these two species at certain localities may have been caused by high trade figures, but a recent review of the conservation status of the two species (Best *et al.* in press) concluded that on on current data it is impossible to determine whether the species are threatened by trade, and further research is required.

Species with exceptionally small ranges

Two Tumbesian endemics are especially vulnerable because of their very small ranges: *Penelope albipennis* and *Atlapetes pallidiceps*. The known ranges of these species do not exceed 1,000 km^2 and they are classic examples of "extinction prone species" (Terborgh 1974), being especially vulnerable to natural disasters, habitat alteration and other human pressures in their tiny ranges. The remaining wild population of *Penelope albipennis* may number less than 250 individuals, whereas *Atlapetes pallidiceps* has not been seen for over 23 years and may already be extinct (Collar *et al.* 1992). The comparatively small ranges of *Pyrrhura orcesi* and *Acestrura berlepschi* expose these two species to increased risks from habitat destruction.

THE RED DATA BOOK LISTING OF THE TUMBESIAN AVIFAUNA

The most authoritative work on the threatened birds of South America is *Threatened birds of the Americas* (Collar *et al.* 1992). The threatened species categories in that work have been updated in *Birds to Watch 2* (Collar *et al.* 1994) using new IUCN criteria. This latter work lists 16 Tumbesian species as globally threatened, with a further six listed as near-threatened (Table 9). These categories are not fixed and both upgrading and downgrading of species should occur as more data become available. They represent the state of knowledge up to 1994. Full accounts for globally threatened species appear in Collar *et al.* (1992). The following section summarizes the most important information on these species, supplemented by distribution maps of all records; and a diagrammatic presentation of the habitat preferences of each species.

Table 9. The threatened and near-threatened Tumbesian bird species.

Species	Status	Threats
Crypturellus transfasciatus	Near-threatened	1, 2
Leucopternis occidentalis	Endangered (2)	1
Ortalis erythroptera	Vulnerable	1, 2
Penelope albipennis	Critical (1)	1, 3
Leptotila ochraceiventris	Vulnerable (2)	1
Aratinga erythrogenys	Near-threatened	1, 4
Pyrrhura orcesi	Vulnerable (10)	1
Brotogeris pyrrhopterus	Near-threatened	1, 4
Acestrura berlepschi	Endangered (2)	1, 3
Synallaxis tithys	Vulnerable (2)	1
Syndactyla ruficollis	Vulnerable (2)	1
Hylocryptus erythrocephalus	Vulnerable (7)	1
Myrmeciza griseiceps	Endangered (2)	1
Onychorhynchus occidentalis	Vulnerable (2)	1
Lathrotriccus griseipectus	Vulnerable (10)	1
Ochthoeca piurae	Near-threatened	1, 3
Tumbezia salvini	Near-threatened	1, 3
Attila torridus	Vulnerable (2)	1
Pachyramphus spodiurus	Near-threatened	1
Phytotoma raimondii	Critical (1)	1, 3
Atlapetes pallidiceps	Critical (1)	1, 3
Carduelis siemiradzkii	Vulnerable (9)	1

Status: species classifications as they appear in Collar *et al.* (1994) are given first, based on new IUCN criteria (Mace and Stuart 1994). Number codes are based on priorities for action in Collar *et al.* (1992): 1. situation critical: action urgent, 2. situation serious: action urgent, 3. situation critical: action urgent when population found, 4. situation terminal: action urgent if population found, 5. situation serious but conservation in progress, 6. situation unclear: action urgent if taxonomic status confirmed, 7. conflicting evidence, possibly urgent, 8. birds perhaps in need if and when found, 9. birds largely unprotected and needing attention, 10. birds with populations only partly protected, 11. birds largely protected, but for which vigilance is needed, 12. birds for which further protection is desirable.
Threats: 1. loss or alteration of habitat, 2. hunting, 3. small range or population, 4. Trade (based on Collar *et al.* 1994)

Sources: Collar *et al.* (1992, 1994).

Figure 44 shows the proportions of the endemic Tumbesian avifauna which is threatened, near-threatened and secure.

The following summaries supplement and update Collar *et al.* (1992). For clarity of reading, and to avoid constant repetition, references have not been included in the species texts. The following have been used in addition to Collar *et al.* (1992): Wiedenfeld *et al.* (1985), Ridgely and Robbins (1988), Robbins and

Figure 44. The conservation status of the birds restricted to the Tumbesian region.

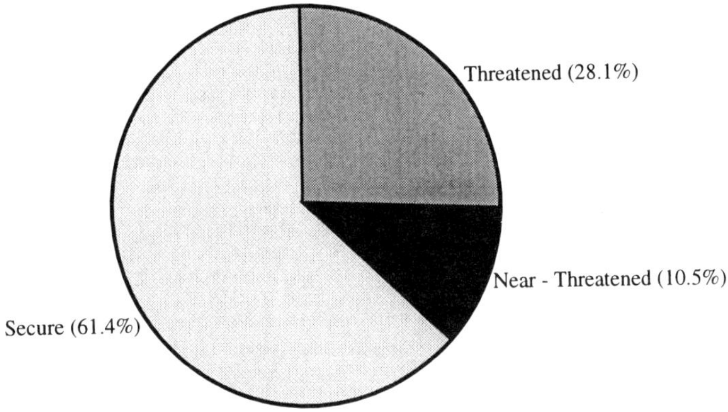

Source: Collar *et al.* (1994).

Ridgely (1990), Best and Clarke (1991), Bloch *et al.* (1991), L. Kiff *in litt.* (1991), T. A. Parker *in litt.* to ICBP (1991), M. B. Robbins *in litt.* 1991, Best *et al.* (1992), P. Coopmans *in litt.* (1992), Parker and Carr (1992), M. Whittingham *in litt.* (1992), Best and Krabbe (1994), Williams and Tobias (1994), Parker *et al.* (1995).

Notes on the species summaries

The following information is presented:

English and Scientific name; Red Data Book listing (see Table 9).

DISTRIBUTION: (Map number), number of traceable localities at which the species has been found; countries and provinces(Ecuador)/departments(Peru).
COORDINATES: N-S and E-W limits.
ALTITUDINAL RANGE: over which each species has been found; for some species the historical range is indicated.
HABITAT PREFERENCES: are listed only if a habitat preferences diagram cannot be produced due to lack of data.
THREATS: listed in order of importance.
PROTECTED AREAS: names of protected areas where the species is known to occur (number of such areas given in parentheses).
SPECIES-SPECIFIC RECOMMENDATIONS: a brief list is given (for further details of conservation action see the 'Conservation recommendations' chapter).

Ornithological effort in the Tumbesian region

The mapped species distributions are potentially dependent on the patterns of observation in the area. Some apparent gaps in the distribution of species are simply due to lack of effort. Figure 45 shows localities in the Tumbesian region which have been surveyed and support Tumbesian endemics up to April 1995.

Figure 45. Localities in Ecuador and northern Peru where restricted-range bird species have been found.

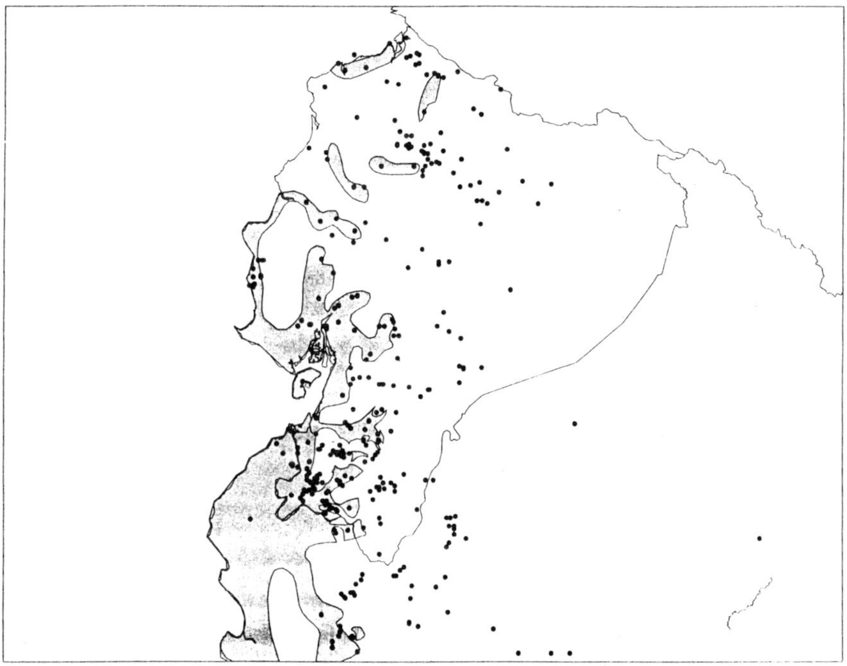

Distribution maps of priority species

Each threatened and near-threatened species has a map of all specimen and reliable sight records. Only localities for which coordinates are available have been mapped.

Habitat preference diagrams

Selected localities in south-west Ecuador (the only region for which a detailed vegetation classification is available: see 'Vegetation' chapter) have been plotted on an altitude/precipitation diagram which separates out the habitat categories (Figure 46). These are principally sites where accurate coordinates, altitude and precipitation figures are available. Different size circles distinguish between well-studied and less well-studied sites. For each species circles are shaded only if it has been found at the site. In this way habitat preferences are immediately obvious. Figure 46 shows that some habitats have received more effort than others; such patterns should be borne in mind by the reader when interpreting the habitat preference ecograms. Well-surveyed sites (large circles) with no record of the species probably means that it was genuinely absent for the duration of the survey. Further details of each site can be found in the site directory on pages 162-174; Figure 90 gives information on the timing of surveys in the region.

THE THREATENED SPECIES

Sixteen Tumbesian endemics are globally threatened (Collar *et al.* 1992). The following pages present distributional and ecological data on them in a standardized form. *Ortalis erythroptera* is treated here as threatened in accordance with Collar *et al.* (1994), but was listed by Collar *et al.* (1992) as near-threatened.

Figure 46. Key for habitat preference diagrams. (see also Figure 32).

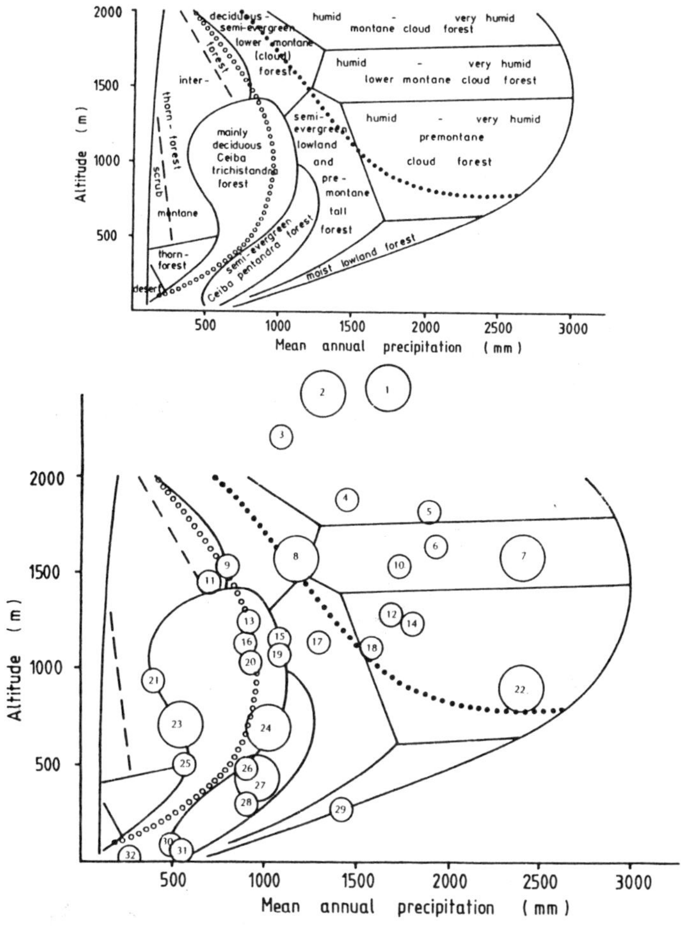

Large circles indicate surveys of 6 days or more have been conducted the site; small circles refer to surveys of 5 days or less.

Sites indicated

Angashcola (2) 4°34'S 79°22'W, **Arenillas (31)** 3°33'S 80°04'W, **Arenillas Military Reserve (32)** 3°33'S 80°03'W, **Buenaventura (22)** 3°40'S 79°40'W, **Campo Verde (24)** 3°51'S 80°11'W, **Cariamanga (3)** 4°20'S 79°33'W, **Catacocha (9)** 4°03'S 79°40'W, **Celica I (4)** 4°09'S 79°50'W, **Celica II (6)** 4°06'S 79°59'W, **8km W. Celica (5)** 4°07'S 79°59'W, **Cruzpampa (17)** 4°10'S 80°01'W, **El Caucho (27)** 3°49'S 80°17'W, **El Empalme (21)** 4°07'S 79°51'W, **El Empalme-Celica (11)** 4°07'S 79°55'W, **Hacienda Yamana (20)** 4°01'S 79°40'W, **Matapalo (30)** 3°41'S 80°12'W, **S. Piñas (16)** 3°40'S 79°43'W, **Puyango (29)** 3°52'S 80°05'W, **Quebrada Las Vegas (12)** 3°59'S 79°59'W, **4km SW. Sabanilla (26)** 4°13'S 80°10'W, **San José de Pozul (10)** 4°07'S 80°03'W, **San Pablo (15)** 3°41'S 79°33'W, **Sozoranga I (13)** 4°18'S 79°47'W, **Sozoranga II (8)** 4°19'S 79°48'W, **Tambo Negro (23)** 4°24'S 79°47'W, **Tierra Colorada (7)** 4°02'S 79°57'W, **Vicentino I (18)** 3°57'S 79°57'W, **Vicentino II (19)** 3°56'S 79°55'W, **Utuana (1)** 4°22'S 79°43'W, **Uzhcurrumi (28)** 3°21'S 79°33'W, **above Uzhcurrumi (14)** 3°23'S 79°32'W, **20km SW. Zapotillo (25)** 4°10'S 80°08'W.

GREY-BACKED HAWK *Leucopternis occidentalis* ENDANGERED

DISTRIBUTION: 34 localities. **Ecuador:** Esmeraldas, Manabí, Guayas, Azuay, El Oro, Loja. **Peru:** Tumbes.
COORDINATES: 0°50'N-4°09'S, 79°06'W-80°46'W
ALTITUDINAL RANGE: sea-level to 2,900 m.
THREATS: deforestation.
PROTECTED AREAS: Cerro Mutiles, Río Palenque Reserve (?), Jauneche Reserve, Machalilla National Park, Cerro Blanco Reserve, Manglares-Churute Ecological Reserve, Arenillas Military Reserve and Tumbes National Forest (8).
SPECIES-SPECIFIC RECOMMENDATIONS: support for Machalilla National Park and Tumbes National Forest.

Figure 47. Distribution of *L. occidentalis*

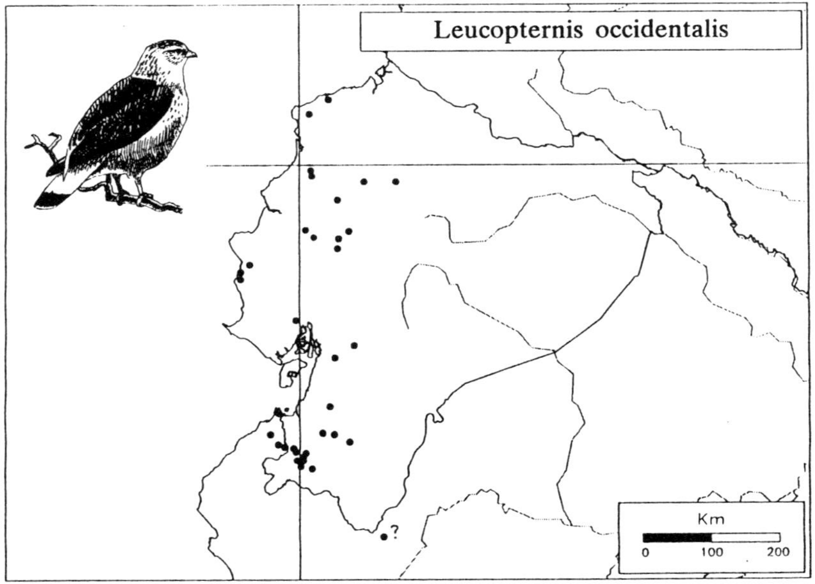

Figure 48. Habitat preferences of *L. occidentalis*

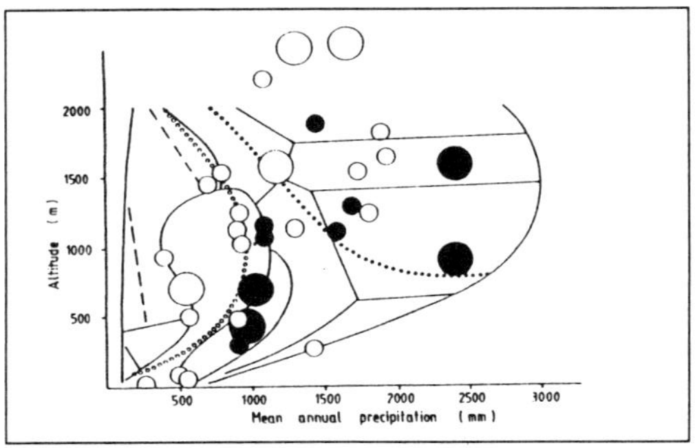

RUFOUS-HEADED CHACHALACA *Ortalis erythroptera* VULNERABLE

DISTRIBUTION: 26 localities. **Ecuador:** Esmeraldas, Pichincha, Manabí, Los Ríos, Azuay, El Oro, Loja. **Peru:** Tumbes.
COORDINATES: 0°50'N-4°24'S, 80°50'W-79°10'W.
ALTITUDINAL RANGE: sea-level to 1,850 m.
THREATS: deforestation, hunting.
PROTECTED AREAS: Cerro Mutiles Reserve, Río Palenque Reserve, Jauneche Reserve, Machalilla National Park, Cerro Blanco Reserve and Tumbes National Forest (6).
SPECIES-SPECIFIC RECOMMENDATIONS: support for Machalilla National Park and Tumbes National Forest.

Figure 49. Distribution of *O. erythroptera*

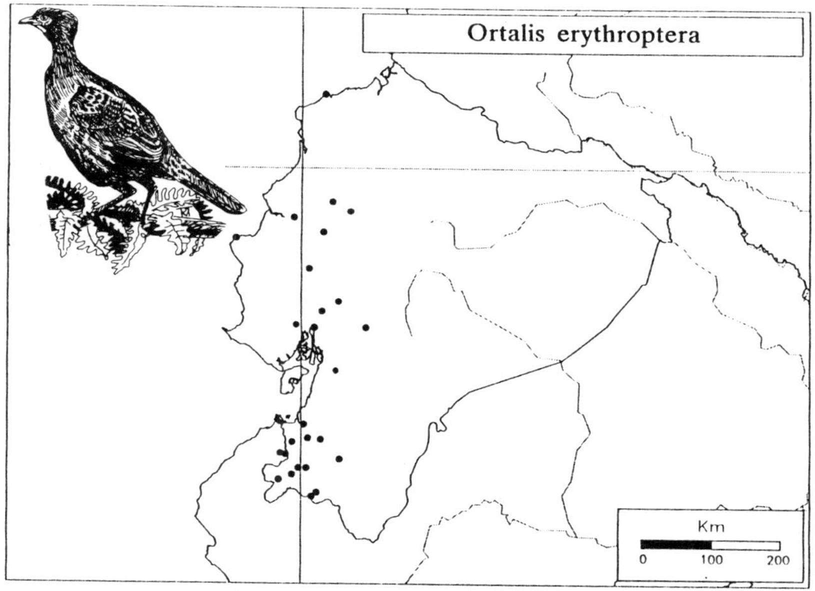

Figure 50. Habitat preferences of *O. erythroptera*

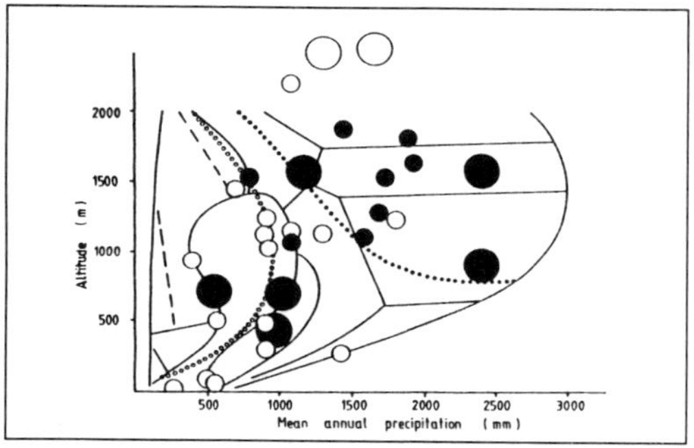

WHITE-WINGED GUAN *Penelope albipennis* CRITICAL

DISTRIBUTION: 16 localities. **Ecuador:** not recorded in current territorial limits. **Peru:** Piura and Lambayeque only.
COORDINATES: currently known only from 5°35'S-5°40'S, 80°20'W-79°24'W.
ALTITUDINAL RANGE: 300-1,200 m (formerly to sea-level).
HABITAT PREFERENCES: dry deciduous forest (formally mangroves).
THREATS: deforestation, hunting, tiny population size.
PROTECTED AREAS: occurs in very small numbers in the tiny Quebrada Negrohuasi Reserve, set up to protect the species.
SPECIES-SPECIFIC RECOMMENDATIONS: (i) search for the species in the rest of south-western Ecuador and north-western Peru to determine the population size, (ii) stringently protect the last strongholds in Peru.

Figure 51. Distribution of *P. albipennis*

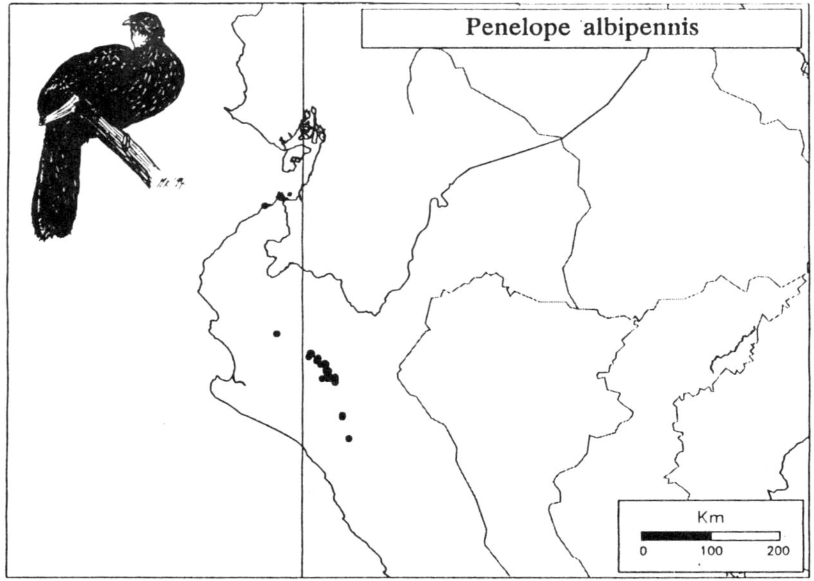

OCHRE-BELLIED DOVE *Leptotila ochraceiventris* VULNERABLE

DISTRIBUTION: 23 localities. **Ecuador:** Manabí, Los Ríos, Guayas, Chimborazo, El Oro, Loja. **Peru:** Tumbes, Piura.
COORDINATES: 0°41′N-5°23′S, 80°40′W-79°00′W.
ALTITUDINAL RANGE: sea-level to 2,625 m.
THREATS: deforestation, understorey disturbance, hunting.
PROTECTED AREAS: Jauneche Reserve, Machalilla National Park, Cerro Blanco, Manglares-Churute Ecological Reserve and Tumbes National Forest (5).
SPECIES-SPECIFIC RECOMMENDATIONS: support for Machalilla National Park and Tumbes National Forest.

Figure 52. Distribution of *L. ochraceiventris*

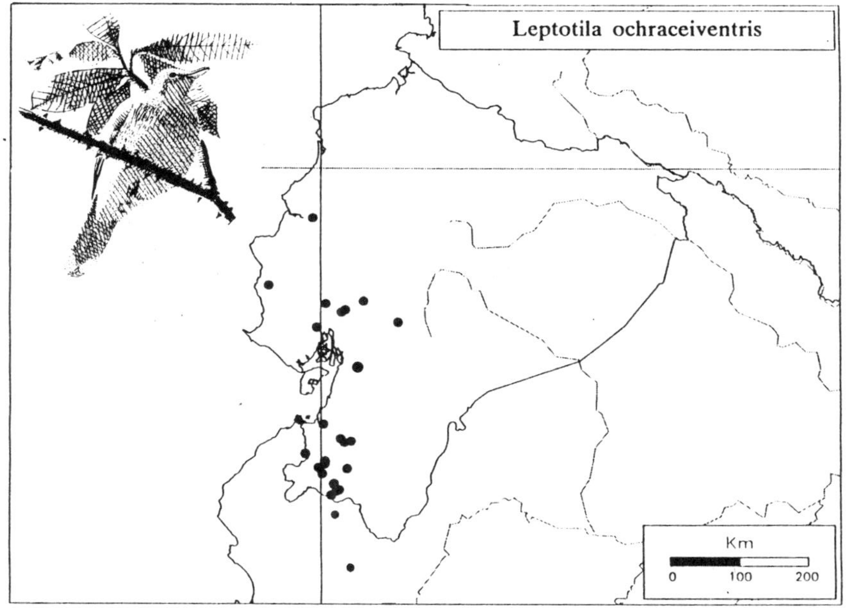

Figure 53 Habitat preferences of *L. ochraceiventris*

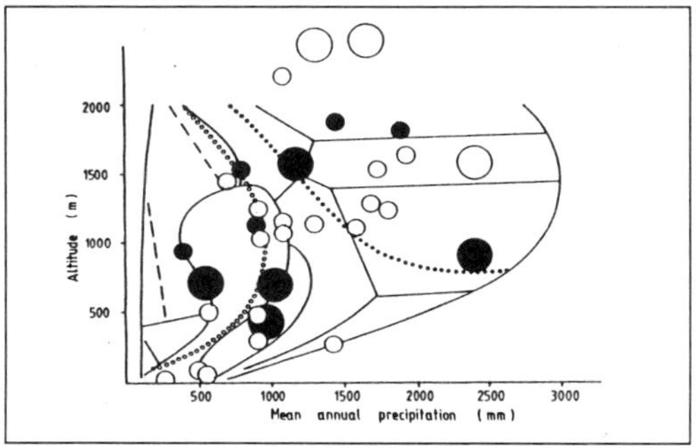

EL ORO PARAKEET *Pyrrhura orcesi* VULNERABLE

DISTRIBUTION: 7 localities. **Ecuador:** Azuay and El Oro only. **Peru:** does not occur.
COORDINATES: 2°30'S-3°39'S, 79°56'W-79°19'W
ALTITUDINAL RANGE: 300-1,300 m.
THREATS: deforestation, exacerbated by the small range of the species.
PROTECTED AREAS: known in none.
SPECIES-SPECIFIC RECOMMENDATIONS: create two protected areas for the species containing two geographically and genetically isolated populations.

Figure 54. Distribution of *P. orcesi*

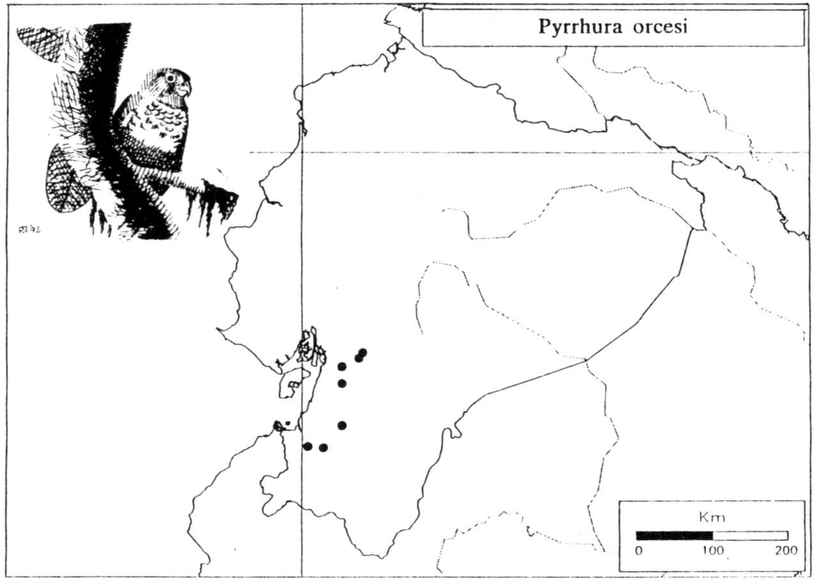

Figure 55. Habitat preferences of *P. orcesi*

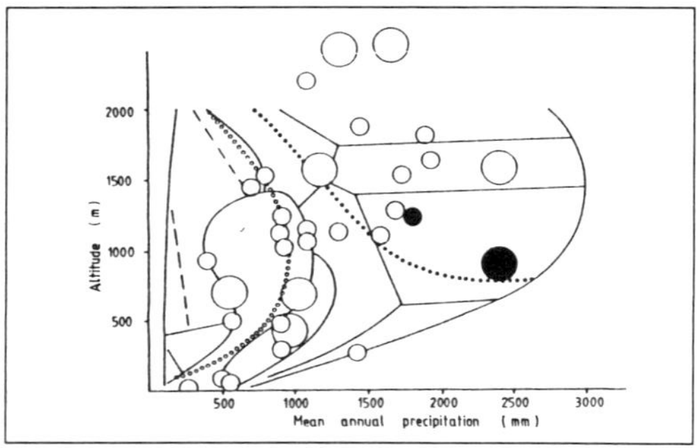

ESMERALDAS WOODSTAR *Acestrura berlepschi* ENDANGERED

DISTRIBUTION: 3 localities. **Ecuador:** Esmeraldas, Manabí and Guayas. **Peru:** no records.
COORDINATES: 0°59′N-1°40′S, 80°45′W-79°42′W.
ALTITUDINAL RANGE: sea-level to 150 m.
HABITAT PREFERENCES: lowland evergreen moist forest.
THREATS: deforestation and understorey disturbance, exacerbated by tiny range.
PROTECTED AREAS: has been found on the edge of the Machalilla National Park.
SPECIES-SPECIFIC RECOMMENDATIONS: (i) promote further protected areas for the species, (ii) determine the size of the remaining population.

Figure 56. Distribution of *A. berlepschi*

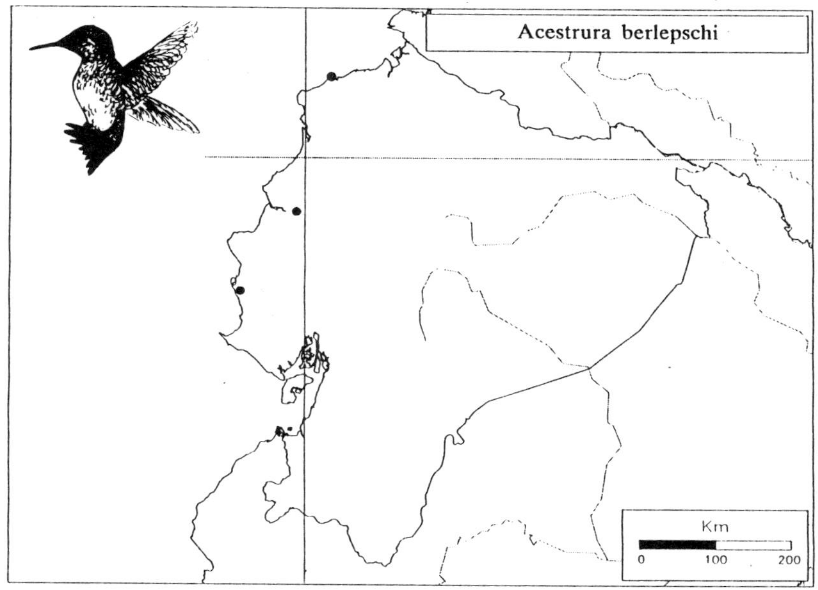

BLACKISH-HEADED SPINETAIL *Synallaxis tithys* VULNERABLE

DISTRIBUTION: 16 localities. **Ecuador:** Manabí, Guayas, El Oro, Loja. **Peru:** Tumbes.
COORDINATES: 1°34′S-4°24′S, 80°40′W-79°50′W.
ALTITUDINAL RANGE: sea-level to 1,000 m.
THREATS: understorey clearance, deforestation.
PROTECTED AREAS: Machalilla National Park, Cerro Blanco Reserve, Arenillas Military Reserve and Tumbes National Forest (4).
SPECIES-SPECIFIC RECOMMENDATIONS: support for Machalilla National Park and Tumbes National Forest.

Figure 57. Distribution of *S. tithys*

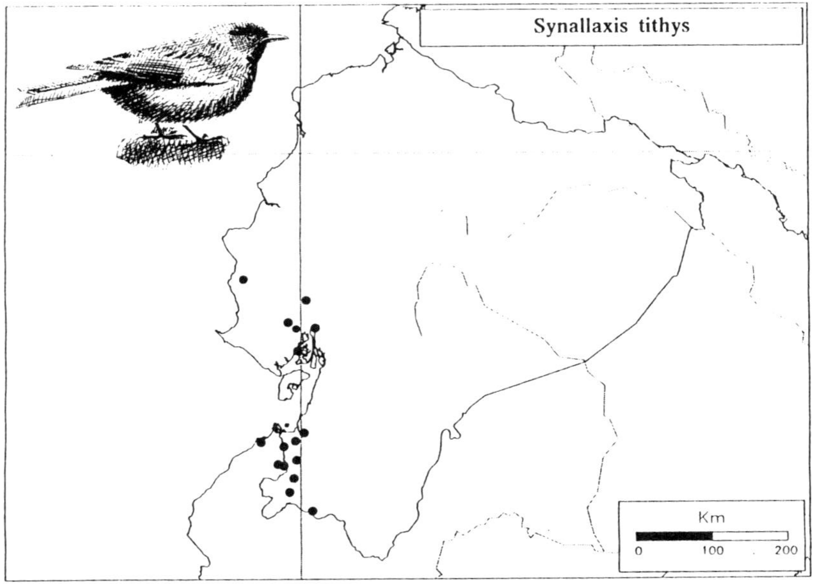

Figure 58. Habitat preferences of *S. tithys*

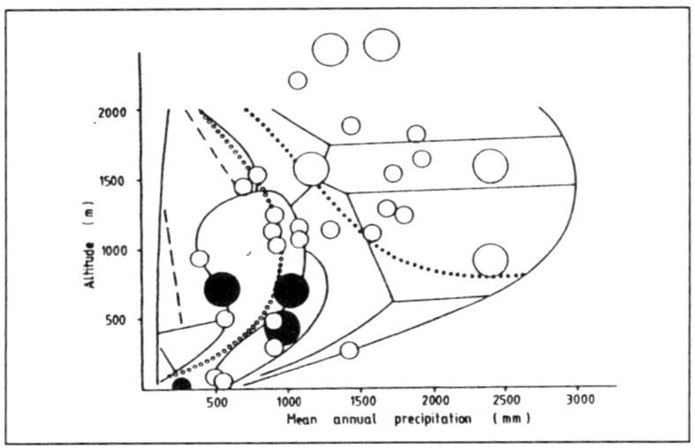

RUFOUS-NECKED FOLIAGE-GLEANER *Syndactyla ruficollis* VULNERABLE

DISTRIBUTION: 30 localities. **Ecuador:** Loja. **Peru:** Tumbes, Piura, Lambayeque, Cajamarca.
COORDINATES: 3°48'S-7°00'S, 80°03'W-78°45'W.
ALTITUDINAL RANGE: 400-2,900 m.
THREATS: understorey disturbance, deforestation.
PROTECTED AREAS: Tumbes National Forest; also occurs in Bosque de Chiñama in Lambayeque Dept., Peru which is being vigorously protected by the local cooperative.
SPECIES-SPECIFIC RECOMMENDATIONS: (i) support for Tumbes National Forest, (ii) promote further protected areas in Loja Province, Ecuador.

Figure 59. Distribution of *S. ruficollis*

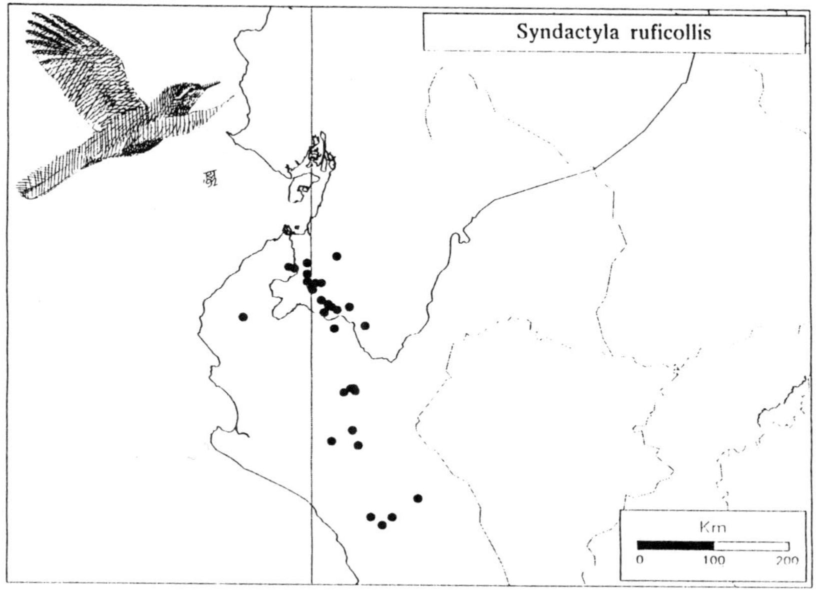

Figure 60. Habitat preferences of *S. ruficollis*

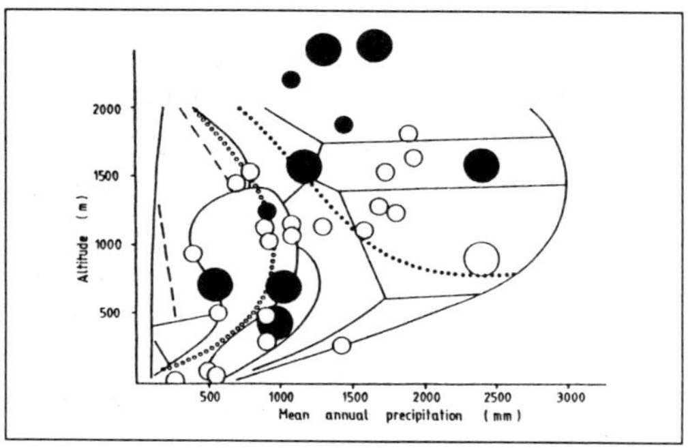

HENNA-HOODED FOLIAGE-GLEANER *Hylocryptus erythrocephalus* VULNERABLE

DISTRIBUTION: 22 localities. **Ecuador:** Manabí, El Oro, Loja. **Peru:** Tumbes, Piura, Lambayeque.
COORDINATES: 1°34′S-5°51′S, 80°40′W-79°37′W.
ALTITUDINAL RANGE: 400-1,750 m.
THREATS: understorey disturbance, deforestation.
PROTECTED AREAS: Machalilla National Park, Cerro Blanco Reserve and Tumbes National Forest (3).
SPECIES-SPECIFIC RECOMMENDATIONS: support for Machalilla National Park and Tumbes National Forest.

Figure 61. Distribution of *H. erythrocephalus*

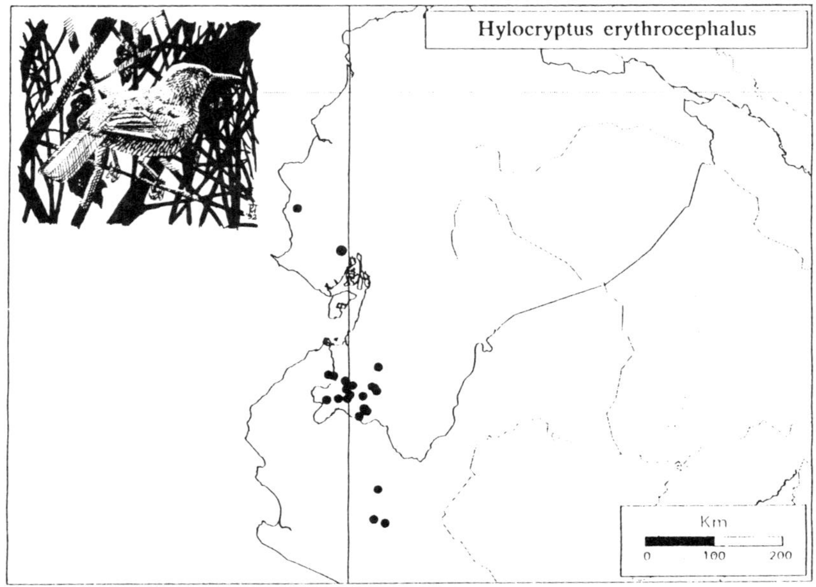

Figure 62. Habitat preferences of *H. erythrocephalus*

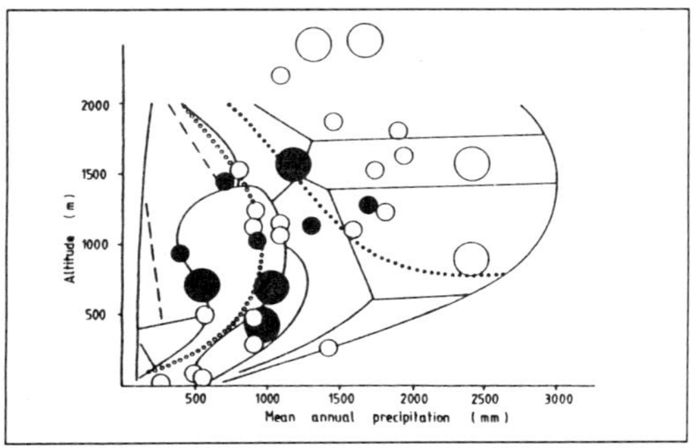

GREY-HEADED ANTBIRD *Myrmeciza griseiceps* ENDANGERED

DISTRIBUTION: 13 localities. **Ecuador:** El Oro and Loja only. **Peru:** Tumbes, Piura.
COORDINATES: 3°35'S-5°51'S, 80°12'W-79°31'W.
ALTITUDINAL RANGE: 600-2,900 m.
THREATS: understorey disturbance, deforestation.
PROTECTED AREAS: Tumbes National Forest.
SPECIES-SPECIFIC RECOMMENDATIONS: (i) create at least one new protected area in Loja Province, Ecuador, (ii) support for Tumbes National Forest.

Figure 63. Distribution of *M. griseiceps*

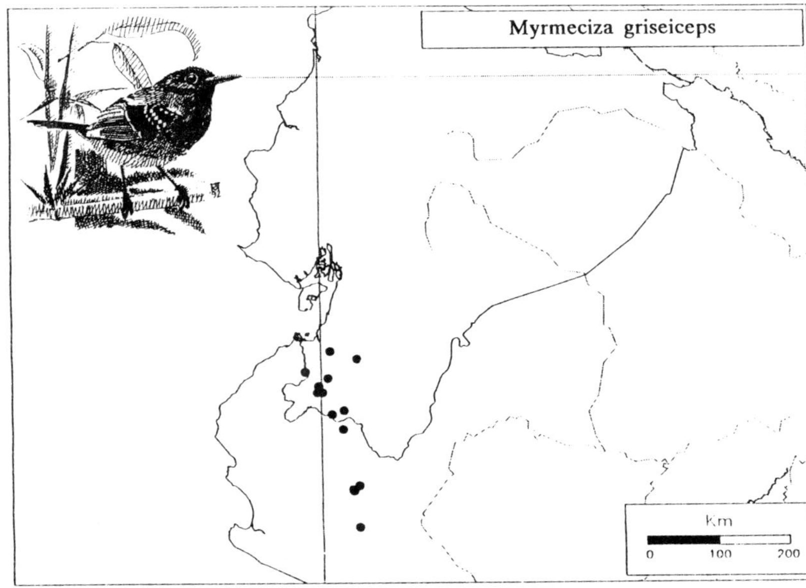

Figure 64. Habitat preferences of *M. griseiceps*

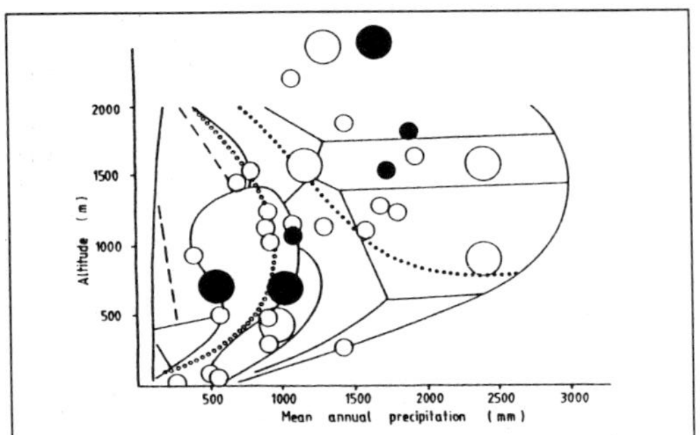

PACIFIC ROYAL-FLYCATCHER *Onychorhynchus occidentalis* VULNERABLE

DISTRIBUTION: 16 localities. **Ecuador:** Esmeraldas, Manabí, Los Ríos, Guayas, Azuay, El Oro. **Peru:** Tumbes.
COORDINATES: 0°59'N-3°50'S, 80°40'W-79°17'W.
ALTITUDINAL RANGE: sea-level to 900 m.
THREATS: deforestation, understorey clearance.
PROTECTED AREAS: Jauneche Reserve, Machalilla National Park, Cerro Blanco Reserve, Manglares-Churute Ecological Reserve and Tumbes National Forest (5).
SPECIES-SPECIFIC RECOMMENDATIONS: (i) support for Machalilla National Park and Tumbes National Forest, (ii) increase the protection of the Jauneche Reserve.

Figure 65. Distribution of *O. occidentalis*

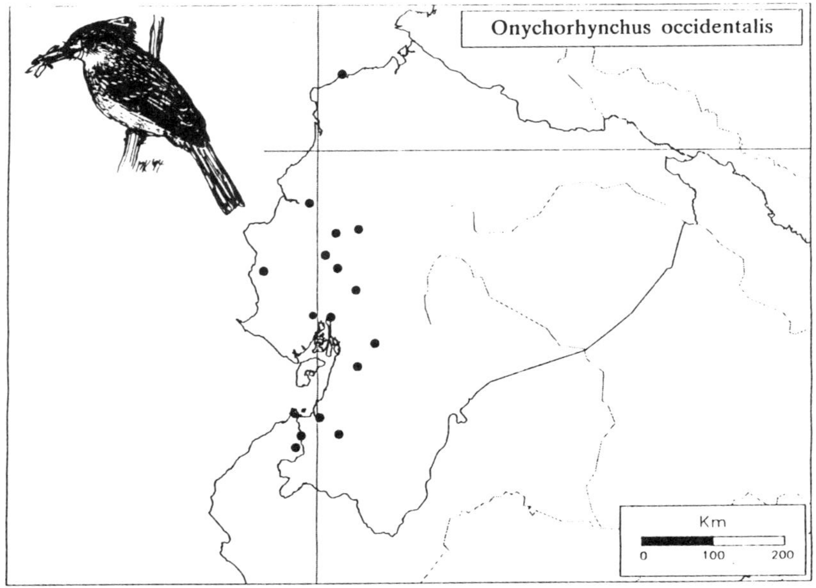

Figure 66. Habitat preferences of *O. occidentalis*

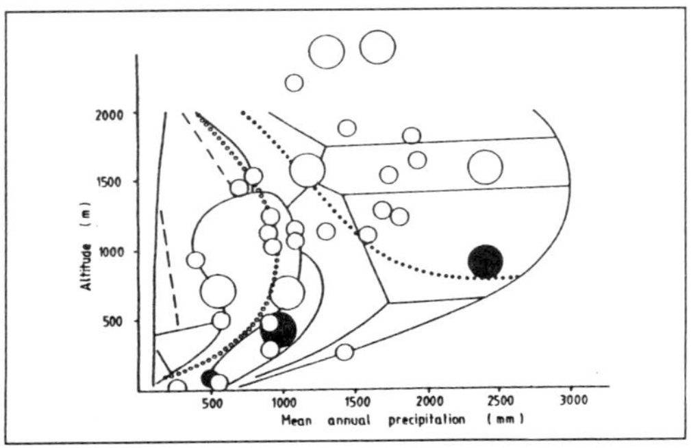

GREY-BREASTED FLYCATCHER *Lathrotriccus griseipectus* VULNERABLE

DISTRIBUTION: 36 localities. **Ecuador:** Esmeraldas, Pichincha, Manabí, Los Ríos, Guayas, Azuay, El Oro, Loja. **Peru:** Tumbes, Piura, Lambayeque, Cajamarca.
COORDINATES: 1°05′N-5°42′S, 80°48′W-78°47′W.
ALTITUDINAL RANGE: sea-level to 1,750 m.
THREATS: understorey disturbance, deforestation.
PROTECTED AREAS: Río Palenque Reserve, Jauneche Reserve, Machalilla National Park, Cerro Blanco Reserve Manglares-Churute Ecological Reserve and Tumbes National Forest (6).
SPECIES-SPECIFIC RECOMMENDATIONS: support for Jauneche Reserve, Machalilla National Park, Manglares-Churute Ecological Reserve and Tumbes National Forest.

Figure 67. Distribution of *L. griseipectus*

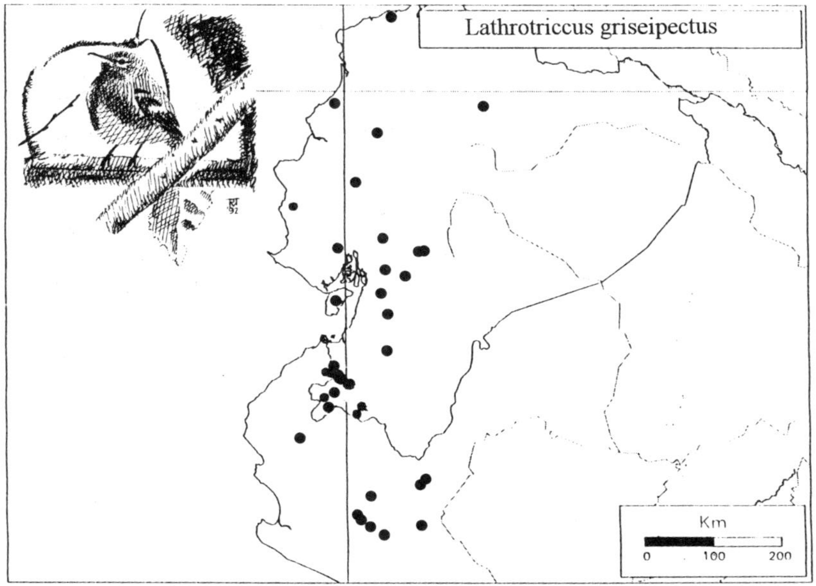

Figure 68. Habitat preferences of *L. griseipectus*

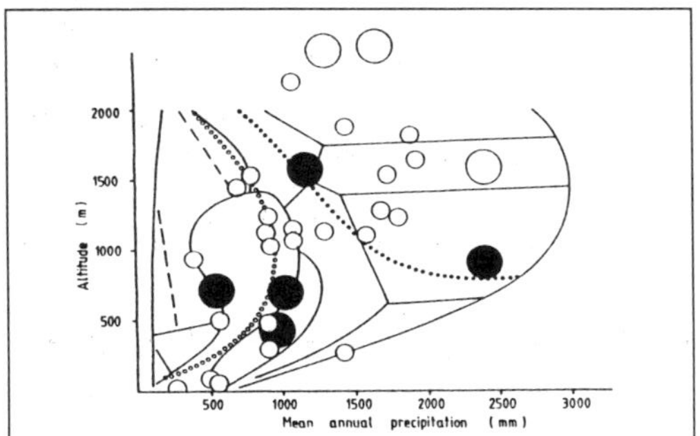

OCHRACEOUS ATTILA *Attila torridus* VULNERABLE

DISTRIBUTION: 27 localities. **Colombia:** Nariño. **Ecuador:** Esmeraldas, Pichincha, Manabí, Los Ríos, Guayas, Azuay, El Oro, Loja. **Peru:** Tumbes.
COORDINATES: 1°29′N-4°06′S, 80°40′W-78°43′W.
ALTITUDINAL RANGE: sea-level to 1 00 m.
THREATS: deforestation, understorey disturbance.
PROTECTED AREAS: Río Palenque Reserve, Jauneche Reserve, Machalilla National Park and Tumbes National Forest (4).
SPECIES-SPECIFIC RECOMMENDATIONS: support for Machalilla National Park and Tumbes National Forest.

Figure 69. Distribution of *A. torridus*

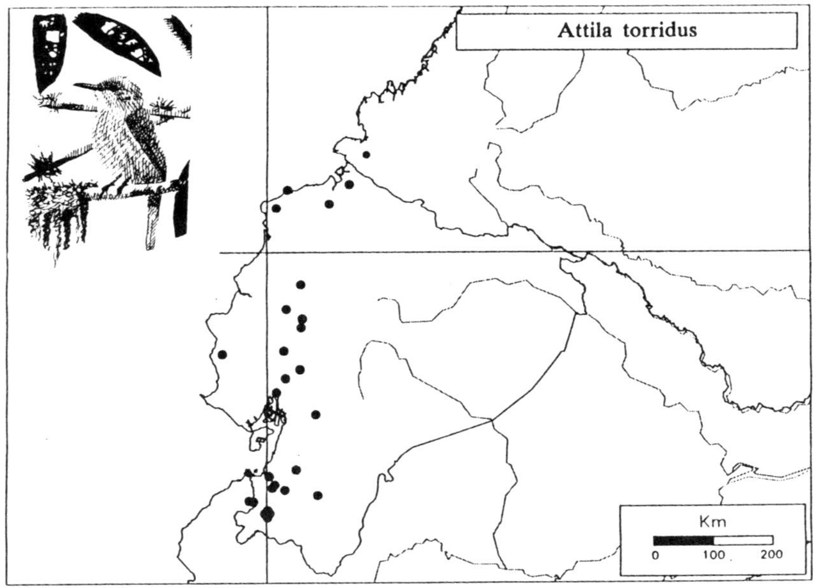

Figure 70. Habitat preferences of *A. torridus*

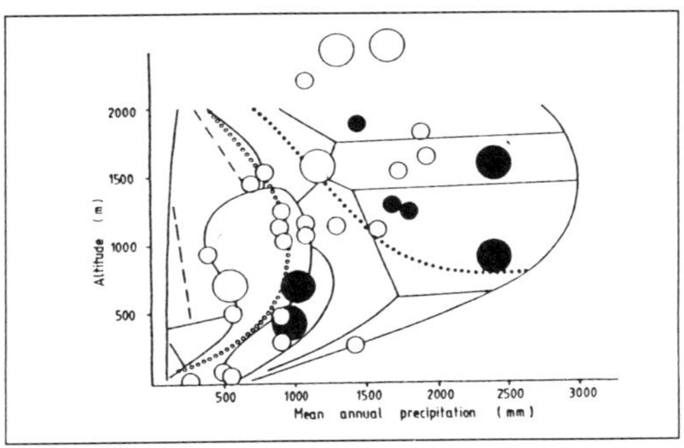

PERUVIAN PLANTCUTTER *Phytotoma raimondii* CRITICAL

DISTRIBUTION: 16 localities. **Ecuador:** not known to occur. **Peru:** Tumbes, Piura, Lambayeque, Libertad, Ancash, Lima.
COORDINATES: 3°34′S-11°30′S, 81°13′W-77°00′W.
ALTITUDINAL RANGE: sea-level to 550 m.
THREATS: deforestation, understorey disturbance.
PROTECTED AREAS: known in none.
SPECIES-SPECIFIC RECOMMENDATIONS: (i) studies of its habitat requirements and threats, (ii) creation of protected areas if required.

Figure 71. Distribution of *P. raimondii*

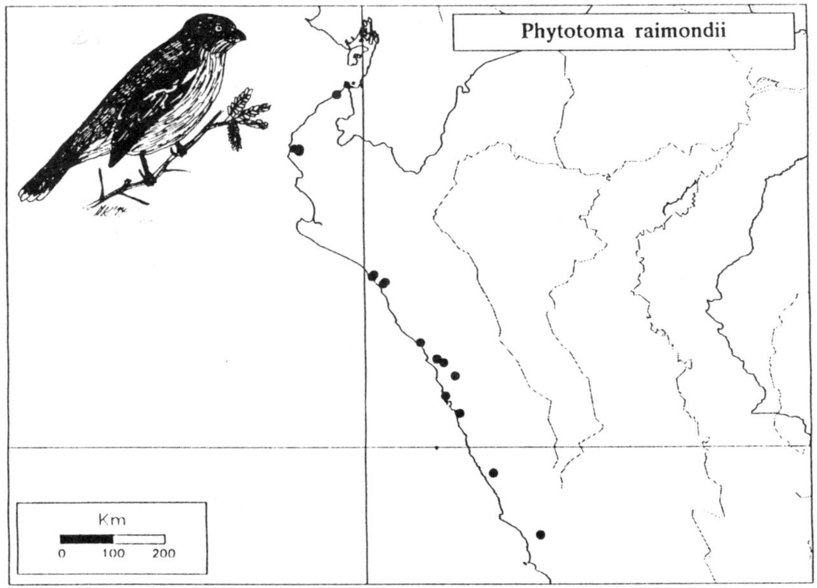

PALE-HEADED BRUSH-FINCH *Atlapetes pallidiceps* CRITICAL

DISTRIBUTION: 6 localities. **Ecuador:** Azuay and Loja only (the single record from Loja is considered erroneous by some observers e.g. M. B. Robbins *in litt.* to ICBP 1992). **Peru:** no records.
COORDINATES: 3°10'S-3°57'S, 79°36'W-79°08'W.
ALTITUDINAL RANGE: 1,500-2,100 m.
HABITAT PREFERENCES: scrub and woodland.
THREATS: deforestation and understorey clearance, exacerbated by tiny range.
PROTECTED AREAS: known in none.
SPECIES-SPECIFIC RECOMMENDATIONS: intensive surveys of the species's former localities, followed by protection of its habitat if it is rediscovered.

Figure 72. Distribution of *A. pallidiceps*

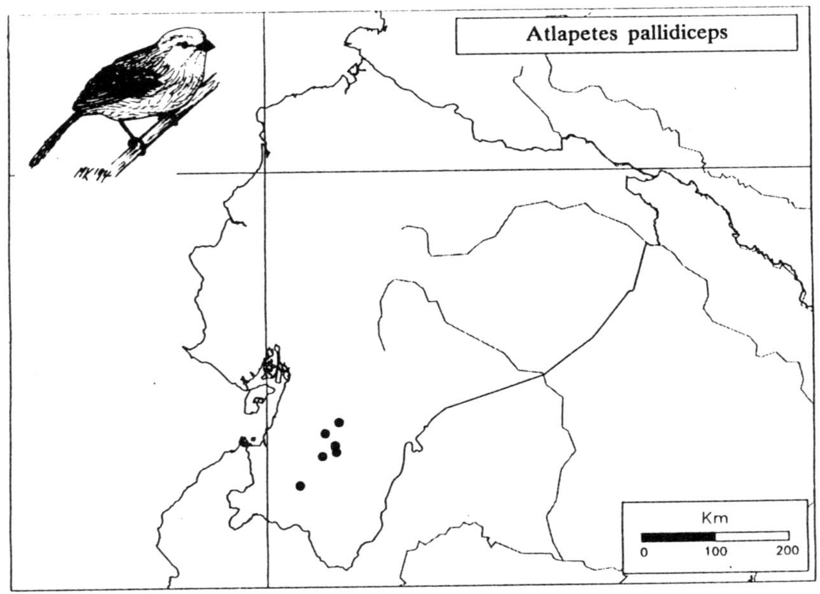

SAFFRON SISKIN *Carduelis siemiradzkii* VULNERABLE

DISTRIBUTION: 13 localities. **Ecuador:** Manabí, Guayas, Loja. **Peru:** Tumbes.
COORDINATES: 0°55′S-4°18′S, 80°45′W-79°44′W.
ALTITUDINAL RANGE: sea-level to 750 m.
THREATS: deforestation?, small range makes the species vulnerable.
PROTECTED AREAS: Machalilla National Park, Cerro Blanco Reserve and Tumbes National Forest (3).
SPECIES-SPECIFIC RECOMMENDATIONS: (i) support for Machalilla National Park and Tumbes National Forest, (ii) research into its habitat requirements.

Figure 73. Distribution of *C. siemiradzkii*

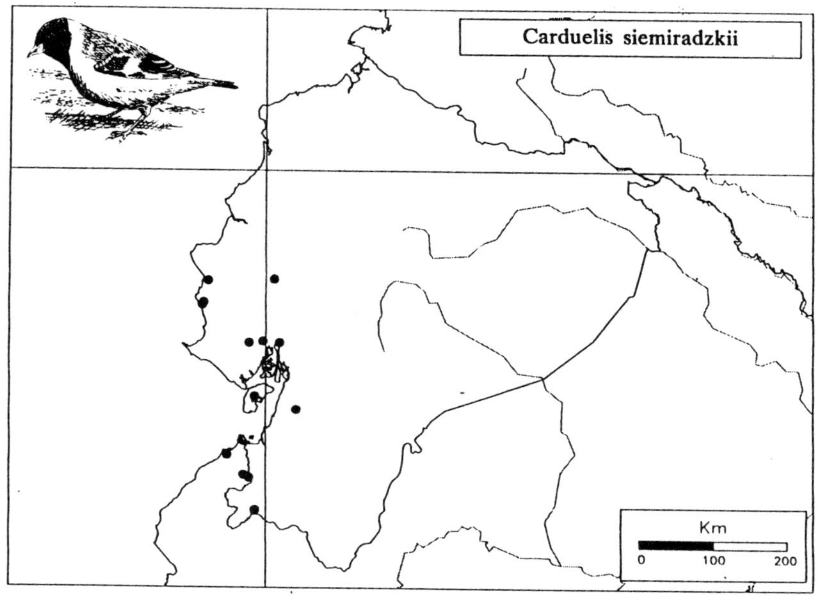

Figure 74. Habitat preferences of *C. siemiradzkii*

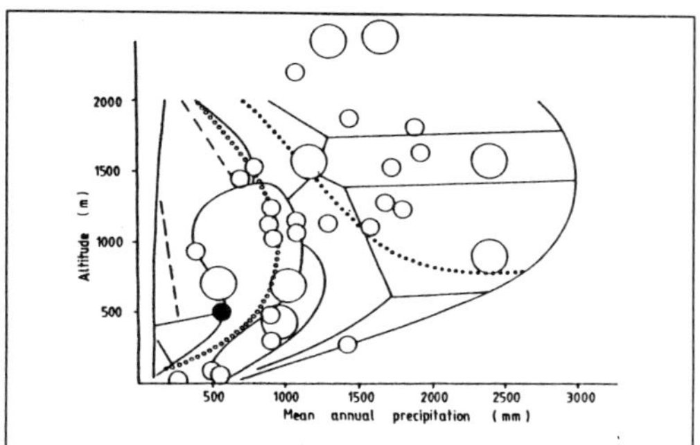

THE NEAR-THREATENED SPECIES

Collar *et al.* (1992) recognize six near-threatened Tumbesian species. One (*Brotogeris pyrrhopterus*) was considered threatened by Collar and Andrew (1988), but subsequent information suggested it is still fairly common in several localities, and may be more tolerant of habitat alteration than originally thought (N. J. Collar verbally 1992). *Pachyramphus spodiurus* is treated here as near-threatened in accordance with Collar *et al.* (1994), although it was listed as threatened in Collar *et al.* (1992). We belive its position is still uncertain, and it may deserve threatened status. *Crypturellus transfasciatus* was given near-threatened status in Collar *et al.* (1992, 1994) and *Aratinga erythrogenys* was listed as near-threatened following recent fieldwork which indicated it may be at risk from habitat destruction, and is also at risk from the cage-bird trade. Such near-threatened species should be carefully monitored, as they may become seriously threatened in the future if current trends in forest clearance and trade continue. In accordance with Collar *et al.* (1992,1994) *Saltator nigriceps* is not treated as near-threatened in this work, although it was listed as such in Collar and Andrew (1988).

The altitudinal ranges of the priority Tumbesian bird species

The altitudinal ranges of threatened and near-threatened Tumbesian birds are quite variable as shown in Figure 85. This should be borne in mind when recommending conservation action in the Tumbesian region.

Figure 75. Known altitudinal limits of the threatened and near-threatened Tumbesian endemics. Note that some species (e.g. *Acestrura berlepschi*) have extremely narrow ranges, whereas others (notably *Leucopternis occidentalis*) occur within much wider limits.

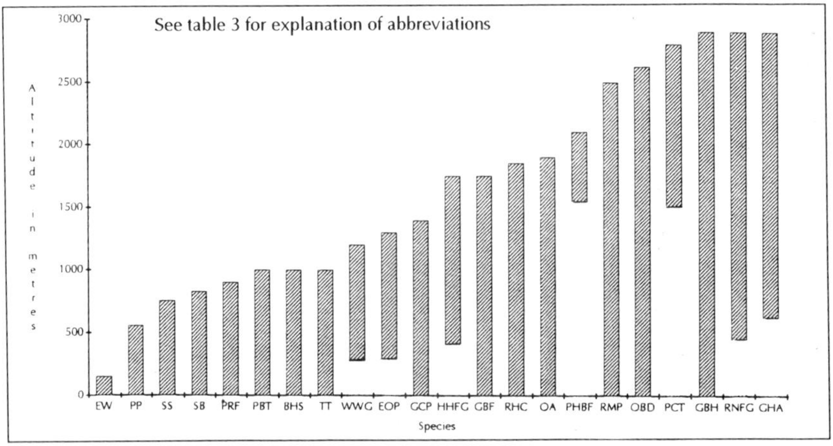

PALE-BROWED TINAMOU *Crypturellus transfasciatus* NEAR-THREATENED

DISTRIBUTION: 16 localities. **Ecuador:** Manabí, Los Ríos, Guayas, El Oro, Loja. **Peru:** Tumbes, Piura, Lambayeque.
COORDINATES: 0°40'S-5°45'S, 80°40'W-79°39'W.
ALTITUDINAL RANGE: sea-level to 1,000 m.
HABITAT PREFERENCES: Figure 76.
THREATS: understorey disturbance, deforestation, hunting.
PROTECTED AREAS: Machalilla National Park, Cerro Blanco Reserve, Arenillas Military Reserve and Tumbes National Forest (4).
SPECIES-SPECIFIC RECOMMENDATIONS: support for Machalilla National Park and Tumbes National Forest.

Figure 76. Distribution of *C. transfasciatus*

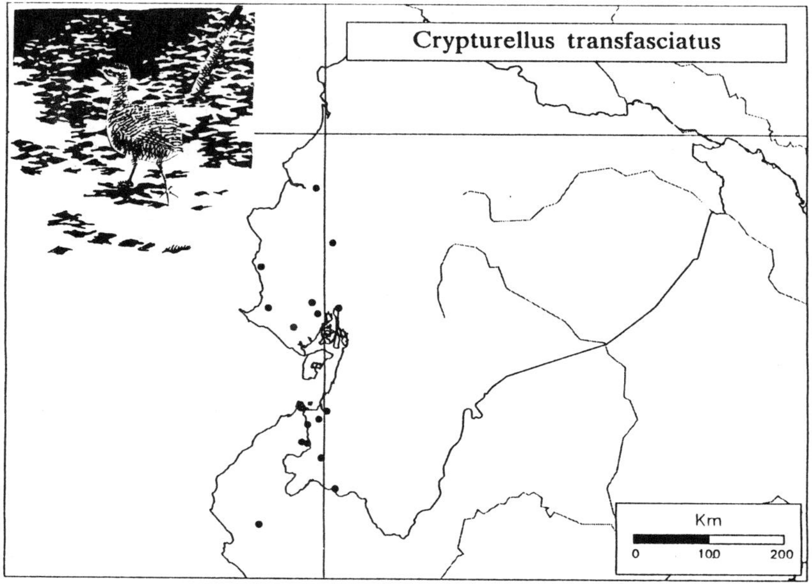

Figure 77. Habitat preferences of *C. transfasciatus*

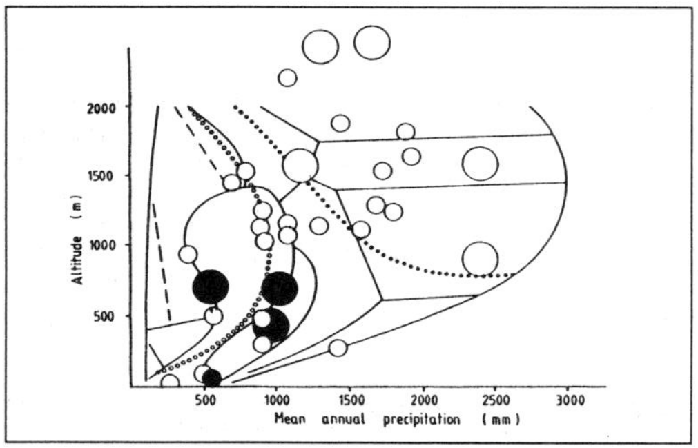

RED-MASKED PARAKEET *Aratinga erythrogenys* NEAR-THREATENED

DISTRIBUTION: 71 traceable localities; occurs in others. **Ecuador:** Esmeraldas, Pichincha, Manabí, Los Ríos, Guayas, Azuay, El Oro, Loja. **Peru:** Tumbes, Piura, Lambayeque.
COORDINATES: 0°32′S-5°59′S, 80°40′W-79°09′W.
ALTITUDINAL RANGE: sea-level to 2,500 m.
HABITAT PREFERENCES: Figure 80.
THREATS: deforestation, bird trade.
PROTECTED AREAS: Cerro Mutiles Reserve, Río Palenque Reserve, Jauneche Reserve, Machalilla National Park, Cerro Blanco Reserve, Manglares-Churute Ecological Reserve, Arenillas Military Reserve, Tumbes National Forest and Cerros de Amotape National Park (9).
SPECIES-SPECIFIC RECOMMENDATIONS: (i) support for Machalilla National Park and Tumbes National Forest, (ii) determine trade sustainability.

Figure 78. Distribution of *A. erythrogenys*

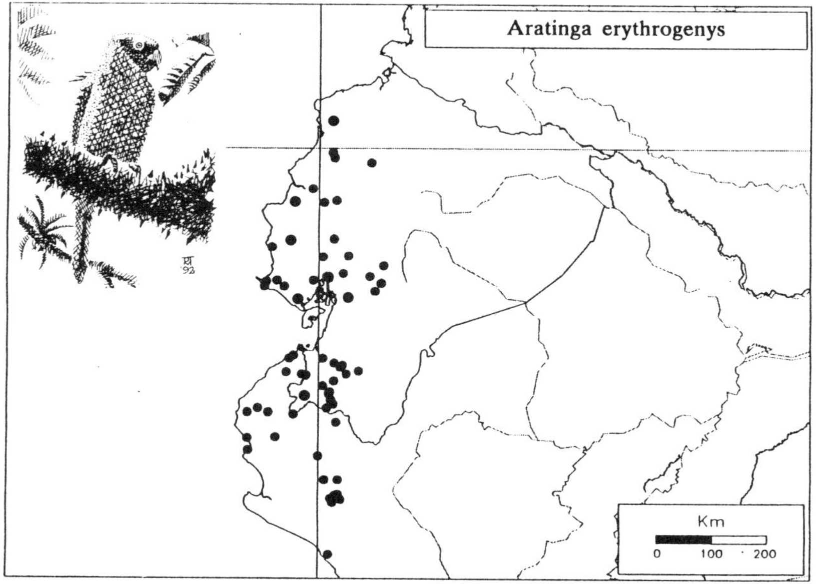

Figure 79. Habitat preferences of *A. erythrogenys*

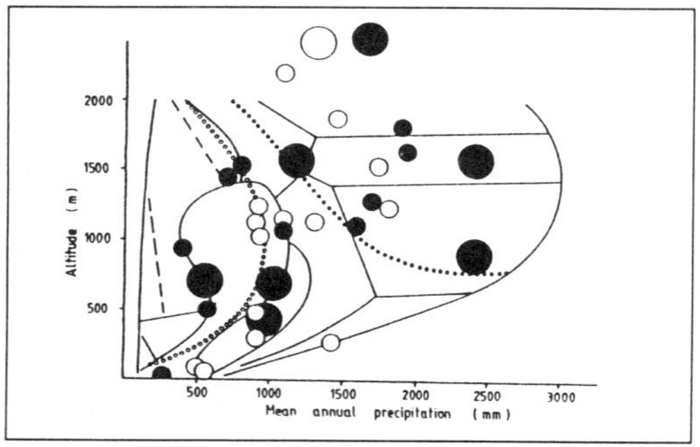

GREY-CHEEKED PARAKEET *Brotogeris pyrrhopterus* NEAR-THREATENED

DISTRIBUTION: 39 traceable localities; occurs in others. **Ecuador:** Manabí, Guayas, Los Ríos, Azuay, El Oro, Loja. **Peru:** Tumbes, Piura.
COORDINATES: 0°32′S-4°26′S, 80°16′W-79°17′W.
ALTITUDINAL RANGE: sea-level to 1,400 m.
HABITAT PREFERENCES: Figure 82.
THREATS: deforestation, bird trade.
PROTECTED AREAS: Cerro Blanco Reserve, Arenillas Military Reserve, Manglares-Churute Ecological Reserve and Tumbes National Forest (4).
SPECIES-SPECIFIC RECOMMENDATIONS: (i) support for Machalilla National Park and Tumbes National Forest, (ii) determine trade sustainability.

Figure 80. Distribution of *B. pyrrhopterus*

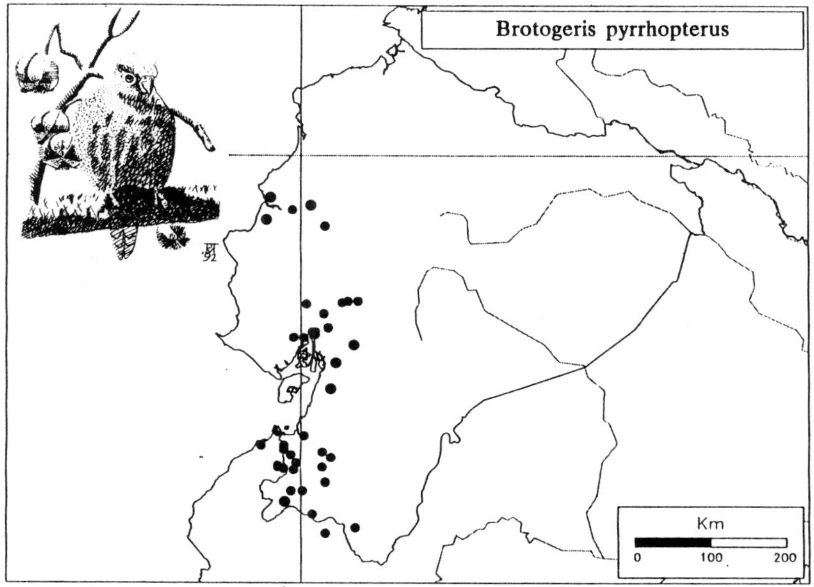

Figure 81. Habitat preferences of *B. pyrrhopterus*

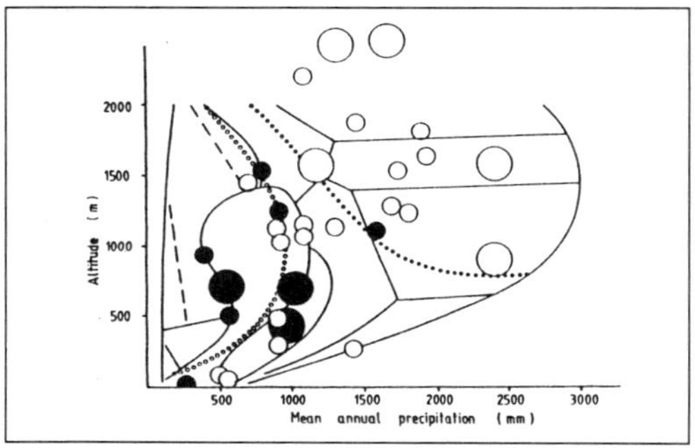

PIURA CHAT-TYRANT *Ochthoeca piurae* NEAR-THREATENED

DISTRIBUTION: 5 localities. **Ecuador:** no records. **Peru:** Piura, Lambayeque, Libertad and Ancash.
COORDINATES: 5°23'S-9°54'S, 77°47'W-79°37'W.
ALTITUDINAL RANGE: 1,500-2,800 m.
HABITAT PREFERENCES: montane scrub, riparian thickets.
THREATS: habitat clearance.
PROTECTED AREAS: known in none.
SPECIES-SPECIFIC RECOMMENDATIONS: (i) carry out intensive surveys to discover additional populations of the species and investigate its habitat requirements, (ii) secure areas known to harbour the species.

Figure 82. Distribution of *O. piurae*

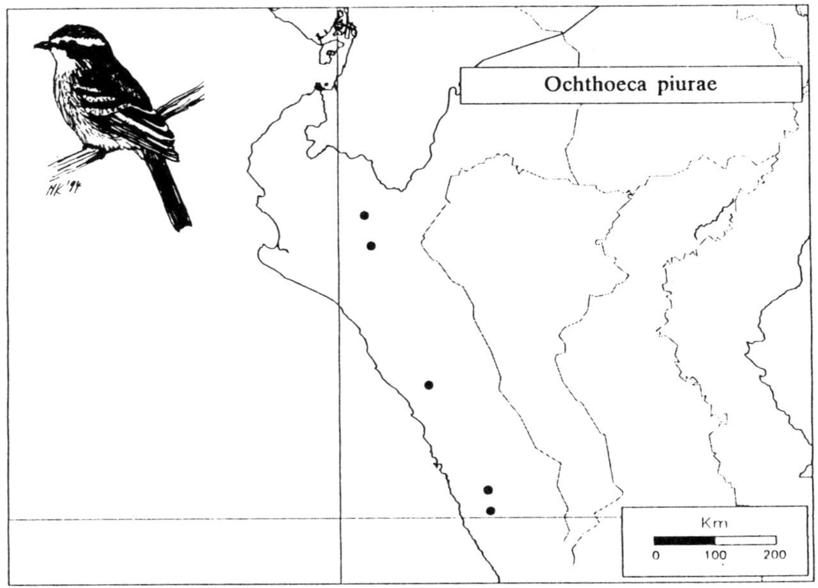

TUMBES TYRANT *Tumbezia salvini* NEAR-THREATENED

DISTRIBUTION: 6 localities. **Peru:** Tumbes, Piura, Lambayeque, Libertad.
COORDINATES: 3°34'S-8°25'S, 78°45'W-80°28'W.
ALTITUDINAL RANGE: sea-level to 1,000 m.
HABITAT PREFERENCES: desert scrub, dry forest, riparian thickets.
THREATS: habitat clearance.
PROTECTED AREAS: known in none.
SPECIES-SPECIFIC RECOMMENDATIONS: (i) search for additional populations and investigate their habitat requirements, (ii) secure areas known to harbour the species.

Figure 83. Distribution of *T. salvini*

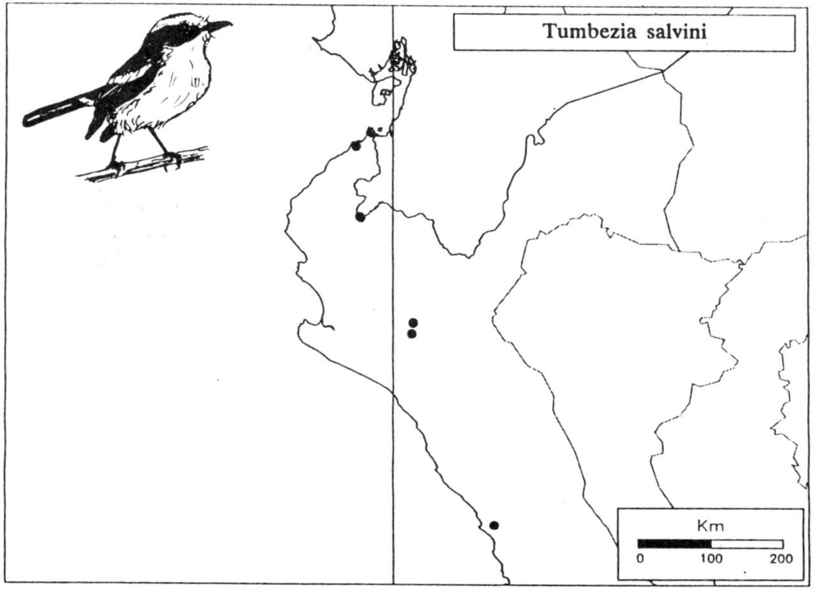

SLATY BECARD *Pachyramphus spodiurus* NEAR-THREATENED

DISTRIBUTION: 29 localities. **Ecuador:** Esmeraldas, Pichincha, Los Ríos, Guayas, Azuay, El Oro, Loja. **Peru:** Tumbes, Piura, Cajamarca, Amazonas.
COORDINATES: 1°00'N-3°44'S, 80°17'W-78°40'W.
ALTITUDINAL RANGE: sea-level to 825 m.
THREATS: understorey disturbance, deforestation.
PROTECTED AREAS: Río Palenque Reserve and Tumbes National Forest (2).
SPECIES-SPECIFIC RECOMMENDATIONS: (i) support for Tumbes National Forest, (ii) protect further areas, (iii) continue to search for the species in Machalilla National Park.

Figure 84. Distribution of *P. spodiurus*

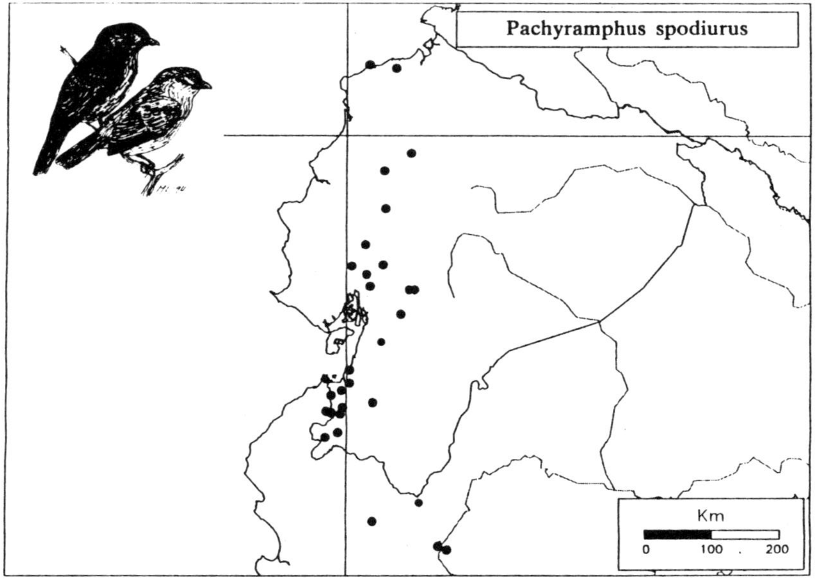

Figure 85. Habitat preferences of *P. spodiurus*

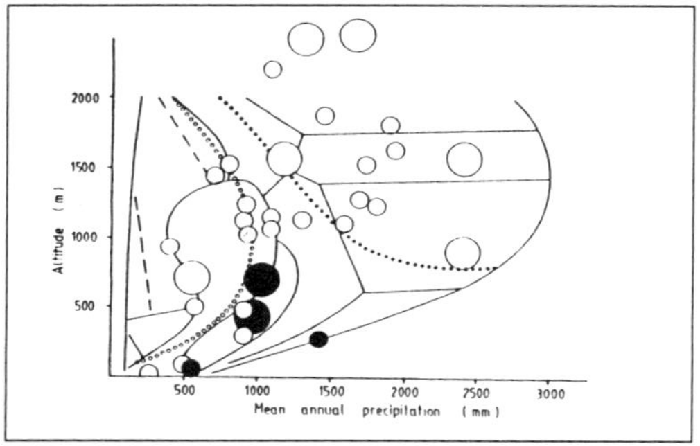

OTHER SPECIES WITH CONSERVATION IMPORTANCE OCCURRING IN THE TUMBESIAN REGION

In addition to the above threatened and near-threatened species which are restricted to or occur largely in the Tumbesian region, 16 other species of conservation importance occur in the region but have larger ranges. These species will be afforded some protection by habitat preservation in the Tumbesian region, but conservation action must also be focused on their distributional centres. These species belong to three groups (within these the threatened species are marked with double asterisks, near-threatened species single asterisks). In addition to these species is the threatened Little Woodstar *Acestrura bombus* which occurs in western and eastern Ecuador and ranges south to central Peru, embracing altitudes from sea-level to 3,050 m (Collar *et al.* 1992). It occurs at several sites in the Tumbesian region, apparently favouring moist and semi-deciduous forests.

Species from the Chocó and Pacific slope Andes EBA

These include two with special conservation importance: Long-wattled Umbrellabird *Cephalopterus penduliger*** and Gorgeted Sunangel *Heliangelus strophianus**. In common with most species from this EBA, these species favour humid cloud-forest which occurs only at scattered sites in the Tumbesian region (e.g. Buenaventura).

Threatened and near-threatened Andean species

Among these are two threatened species from the South Central Andean EBA, *Penelope barbata*** and Rusty-faced Parrot *Hapalopsittaca pyrrhops***. A more wide-ranging but local Andean species is Black-and-chestnut Eagle *Oroaetus isidori**. These three species occur mostly at altitudes of over 2,500 m and favour humid forest, only venturing into lower areas where the forest is very humid. Neblina Metaltail *Metallura odomae** has a restricted range in southern Ecuador and northern Peru, and occurs along the eastern edge of the Tumbesian region at a few humid forest sites above 2,500 m, such as Angashcola, Loja Province, Ecuador. Golden-plumed Parakeet *Leptosittaca branickii*** occurs in humid temperate forest at scattered sites in the Andes of Colombia, Ecuador and Peru. It occurs at a few sites in the Tumbesian region which lie within its preferred altitudinal range (2,500-3,500 m) and possess humid forest. Further Andean species found in at least one site on the eastern edge of the Tumbesian region include Imperial Snipe *Gallinago imperialis**, Buff-fronted Owl *Aegolius harrisii**, Peruvian Antpitta *Grallaricula peruviana**, Orange-banded Flycatcher *Myiophobus lintoni**, Grey-winged Inca-Finch *Incaspiza ortizi** and Masked Mountain-Tanager *Buthraupis wetmorei***.

Three near-threatened premontane species

These are wider-ranging species: Solitary Eagle *Harpyhaliaetus solitarius**, Fasciated Tiger-heron *Tigrisoma fasciatum** and Scaled Fruiteater *Ampelioides tschudii**.

DIRECTORY OF PRIORITY SITES FOR BIRD CONSERVATION IN THE TUMBESIAN REGION

This section presents, in standardized form, information on all sites in the Tumbesian region at which two or more threatened or near-threatened Tumbesian endemics have been recorded since 1970. The 'Conservation recommendations' chapter uses this information to identify the most important Tumbesian sites for conserving of its avifauna. The information presented for each site is listed below. The geographical positions of the localities are shown in Figure 86. Sites 1-24 are in Ecuador; sites 25-30 in Peru.

NAME: the most commonly used name is given first; other names are given in parentheses. The province/department is given on the title line.
COORDINATES: if the site is a large one then the central coordinates only are given.
ALTITUDE and DESCRIPTION OF SITE: if the site falls within the area of the vegetation classification presented earlier is given, otherwise a broader habitat classification is given based on the references.
FOREST EXTENT: estimates refer to the last survey of each site, not necessarily the 1995 situation.
SURVEY DATES: these concentrate on recent efforts, i.e. those after 1970.
ORNITHOLOGICAL DATA: species totals are given only if detailed lists have been published or unpublished lists made available. Both Tumbesian and non-Tumbesian threatened and near-threatened species are listed.
PROTECTED STATUS and THREATS: appropriate details are given.
REFERENCES: Initials of surveyors/institutions mentioned correspond to: (AB) A. Brosset; (ANSP) Academy of Natural Sciences, Philadelphia; (BJB *et al.*) B. J. Best, A. L. Broom, M. Checker, J. W. Duckworth, M. Kessler, R. Thewlis; (BJB *et al.**) B. J. Best, C. T. Clarke, M. Checker, A. McNab; (CTC) C. T. Clarke; (DAW *et al.*) D. A. Wiedenfeld, T. S. Schulenberg, M. B. Robbins; (EK) E. Krabbe; (HB *et al.*) H. Bloch, M. K. Poulsen, C. Rahbek and J. F. Rasmussen; (FL) F. Lambert; (LSUMZ) Louisiana State University Museum of Zoology; (MC *et al.*) M. Checker, R. Thewlis, W. Duckworth and M. Kessler; (MCZ) Museum of Comparative Zoology, Harvard; (MK) M. Kessler; (NK) N. Krabbe; (PC) P. Coopmans; (TAP) the late T. A. Parker; (RSR) R. S. Ridgely; (TM) T. Meyers; (TSS) T. S. Schulenberg; (MW) M. Whittingham; (RSRW and JAT) R. S. R. Williams and J. A. Tobias; (WFVZ) Western Foundation of Vertebrate Zoology, Los Angeles.

Figure 86. Locations of important ornithological sites in the Tumbesian region.
Numbers refer to a site's listing in the directory.

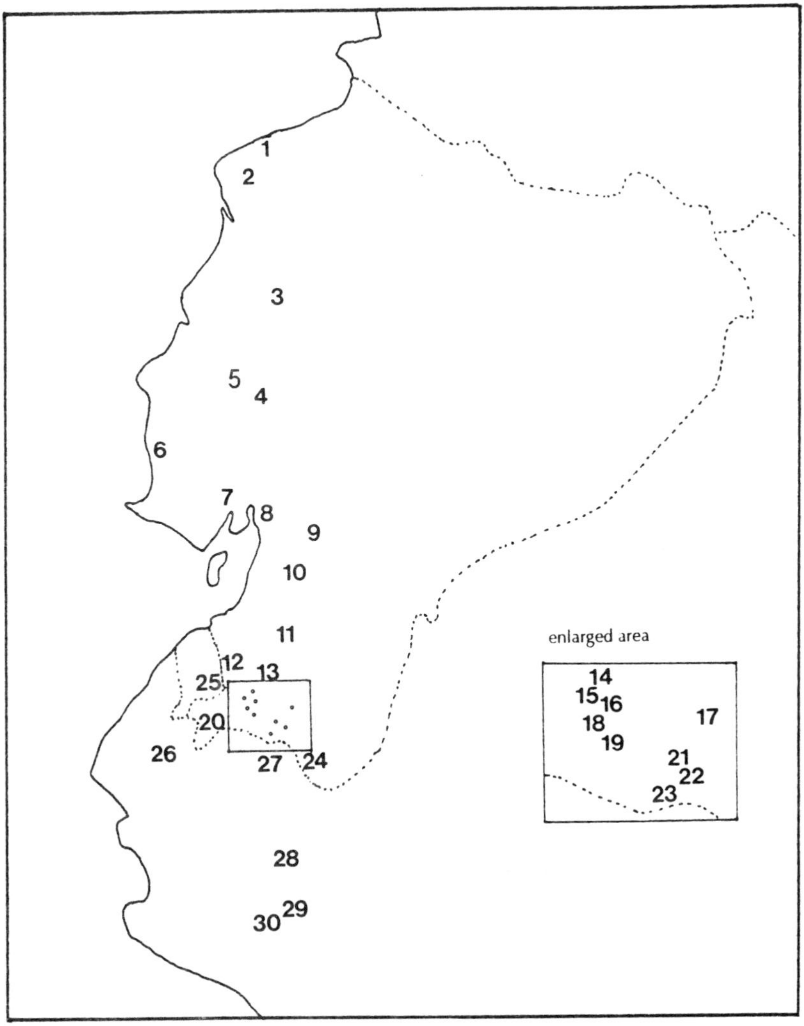

1. Cerro Mutiles — Esmeraldas

(Cerro San Mateo, Reserva Jardín Tropical "Luis Vargas Torres").

COORDINATES: 0°54'N 79°57'W.
ALTITUDE: 60-300 m.
DESCRIPTION: a small area of moist, semi-evergreen forest on a ridge SE of Esmeraldas, N of río Esmeraldas.
FOREST EXTENT: not known (but said to be "small").
SURVEY DATES: 2-4 Feb 1991 (TAP); 3 observer days.
ORNITHOLOGICAL DATA: 136 species recorded including two threatened species (*Leucopternis occidentalis* and *Ortalis erythroptera*).
PROTECTED STATUS: owned by the National University of Esmeraldas.
THREATS: logging inside the reserve.
REFERENCES: Parker and Carr (1992).

2. Cabeceras de Bilsa — Esmeraldas

COORDINATES: 0°37'N, 79°51'W.
ALTITUDE: 100-300 m.
DESCRIPTION: part of a large block of wet forest in near-pristine condition.
FOREST EXTENT: 20,000 ha.
SURVEY DATES: 26-31 Jan 1991.
ORNITHOLOGICAL DATA: 158 species recorded, including four threatened species (*Leucopternis occidentalis, Ortalis erythroptera, Cephalopterus penduliger* and *Attila torridus*).
PROTECTED STATUS: none, but an Ecuadorian conservation group, Fundación Jatun Sacha, is attempting to buy an 800 ha area of forest nearby (see note).
THREATS: further clearance by settlers.
REFERENCES: Parker and Carr (1992).
Note: the Jatun Sacha Bilsa area at 0°22'N, 79°45'W supports the above species and *Pachyramphus spodiurus* and is apparently earmarked for protection by a local conservation group (McColm *et al.* 1994, R. Clay *in litt.* to BirdLife International 1995).

3. Río Palenque Science Centre — Pichincha

COORDINATES: 0°30'S 79°30'W.
ALTITUDE: 200 m.
DESCRIPTION: one of the last few remaining areas of tropical moist forest in western Ecuador. This isolated forest patch is surrounded by agricultural land.
FOREST EXTENT: 167 ha (only 87 ha of forest).
SURVEY DATES: numerous dates throughout the year.
ORNITHOLOGICAL DATA: 355 species recorded, including seven threatened species (*Leucopternis occidentalis* [but not since the 1970s], *Ortalis erythroptera, Acestrura bombus, Cephalopterus penduliger, Onycorhynchus occidentalis, Lathrotriccus griseipectus* and

cont.

Río Palenque Science Centre (cont.)

Dacnis berlepschi) and two near-threatened species (*Aratinga erythrogenys* and *Pachyramphus spodiurus* [only one recent record]).
PROTECTED STATUS: a biological station owned by the University of Miami.
THREATS: Agricultural encroachment.
REFERENCES: Dodson and Gentry (1978), Leck (1979), Leck *et al.* (1980), Parker and Carr (1992).

4. JAUNECHE — Los Ríos

COORDINATES: 1°20'S, 79°35'W.
ALTITUDE: 50-70 m.
DESCRIPTION: one of the last tropical moist forests in western Ecuador, containing some seasonally inundated forest in the eastern part.
FOREST EXTENT: 138 ha.
SURVEY DATES: 6-9 July 1991 (TAP), 31 Aug-3 Sept (PC), 2-4 Oct (RSRW and JAT).
ORNITHOLOGICAL DATA: five threatened species (*Ortalis erythroptera, Leptotila ochraceiventris, Onychorhynchus occidentalis, Lathrotriccus griseipectus* (up to 3 pairs/ha) and *Attila torridus*) and two near-threatened species (*Crypturellus transfasciatus* and *Aratinga erythrogenys*).
PROTECTED STATUS: a biological station owned by the University of Guayaquil.
THREATS: apparently well protected.
REFERENCES: Dodson *et al.* (1985), P. Coopmans *in litt.* (1992), Parker and Carr (1992), Williams and Tobias (1994).

5. HACIENDA PACARITAMBO — Los Ríos

COORDINATES: 1°02'S, 79°29'W.
ALTITUDE: not known.
DESCRIPTION: moist forest patches and agricultural land.
FOREST EXTENT: no patches exceed 10 ha.
SURVEY DATES: 18-28 May, 15-20 June 1962; 10-15 Feb 1963 (AB).
ORNITHOLOGICAL DATA: 125 species recorded, including three threatened species (*Leucopternis occidentalis, Onychorhynchus occidentalis* and *Attila torridus*) and two near-threatened species (*Brotogeris pyrrhopterus* and *Pachyramphus spodiurus*).
PROTECTED STATUS: none.
THREATS: not known.
REFERENCES: Brosset (1964), Vuilleumier (1978).

6. Machalilla National Park — Manabí

COORDINATES: 1°36'S, 80°42'W.
ALTITUDE: sea-level to 800 m.
DESCRIPTION: several forested hills, secondary forest and some settlements; forest understorey largely undisturbed. Observations so far concentrated on Cerro San Sebastian.
FOREST EXTENT: 15,000 ha.
SURVEY DATES: 10-13 July 1978 (RSR), 18-24 Jan 1991 (TAP), 31 July-10 Aug 1991 (ANSP).
ORNITHOLOGICAL DATA: 214 species recorded, including 11 threatened species (*Leucopternis occidentalis, Ortalis erythroptera, Leptotila ochraceiventris, Acestrura bombus, Acestrura berlepschi, Synallaxis tithys, Hylocryptus erythrocephalus, Onychorhynchus occidentalis, Lathrotriccus griseipectus, Attila torridus* and *Carduelis siemiradzkii*) and two near-threatened species (*Crypturellus transfasciatus* and *Aratinga erythrogenys*).
PROTECTED STATUS: Ecuadorian government-designated National Park.
THREATS: animal grazing, selective logging, settlers within the park.
REFERENCES: MacBryde (1987), Ridgely (1991a), Parker and Carr (1992).

7. Cerro Blanco — Guayas

(Cerro Azul, "Cemento Nacional")

COORDINATES: 2°10'S, 80°02'W.
ALTITUDE: 100-420 m.
DESCRIPTION: a semi-evergreen forest on the edge of the Chongón-Colonche Hills. Formerly owned by the Ecuadorian National Cement Company, was managed by Fundación Natura from 1990 to 1993, now administered by Fundación Pro-Bosque.
FOREST EXTENT: 2,000 ha.
SURVEY DATES: 17 Jan and 15-20 July 1991 (TAP) also regular visits from August 1992 to May 1993 by K. S. Berg and several surveys by PC and R. Jones.
ORNITHOLOGICAL DATA: At least 143 species recorded, including eight threatened species (*Leucopternis occidentalis, Ortalis erythroptera, Leptotila ochraceiventris, Synallaxis tithys, Hylocryptus erythrocephalus, Onychorhynchus occidentalis, Lathrotriccus griseipectus* and *Carduelis siemiradzkii*) and three near-threatened species (*Crypturellus transfasciatus, Aratinga erythrogenys* and *Brotogeris pyrrhopterus*).
PROTECTED STATUS: private reserve.
THREATS: apparently well protected, but illegal settlers have recently arrived on the north side of the reserve.
REFERENCES: P. J. Greenfield *in litt.* to ICBP (1989), Parker and Carr (1992), K. S. Berg *in litt.* (1993), R. Phillips *in litt.* (1994).

8. Manglares-Churute Ecological Reserve — Guayas

COORDINATES: 2°25'S, 79°37'S.
ALTITUDE: sea-level to 900 m.
DESCRIPTION: forest, woodland and marsh.
FOREST EXTENT: reserve extends to 35,000 ha; forest extent not known.
SURVEY DATES: 24-26 Jan 1991 (RSR), also recent surveys by K. S. Berg and N. Hilgert.
ORNITHOLOGICAL DATA: five threatened species (*Leucopternis occidentalis, Ortalis erythroptera, Leptotila ochraceiventris, Onychorhynchus occidentalis* and *Lathrotriccus griseipectus*) and two near-threatened species (*Aratinga erythrogenys* and *Brotogeris pyrrhopterus*).
PROTECTED STATUS: Ecuadorian government designated Ecological Reserve.
THREATS: selective logging, settlement and domestic animals grazing the understorey.
REFERENCES: R. S. Ridgely *in litt.* (1992).

9. Manta Real — Azuay

COORDINATES: 2°34'S, 79°21'W.
ALTITUDE: 250-1,100 m.
DESCRIPTION: tropical moist forest with some agricultural land in the lower part near the town. A road was built through the area in early 1993.
FOREST EXTENT: on the edge of the 28,000 ha Molleturo Protected Forest.
SURVEY DATES: 26-27 Jan 1991 (ANSP), 10-17 July 1991 (TAP), 15-21 Aug 1991 (ANSP). Also regular surveys since November 1991 by K. S. Berg, stationed at Manta Real.
ORNITHOLOGICAL DATA: at least 130 species recorded, including seven threatened species (*Leucopternis occidentalis, Ortalis erythroptera, Pyrrhura orcesi, Cephalopterus penduliger, Onychorhynchus occidentalis, Lathrotriccus griseipectus* and *Attila torridus*) and four near-threatened species (*Aratinga erythrogenys, Brotogeris pyrrhopterus, Ampelioides tschudii* and *Pachyramphus spodiurus*).
PROTECTED STATUS: none at present, but apparently designated for protection by the Corporación Ornitologica del Ecuador (BirdLife International's counterpart in Ecuador). It has carried out community development and conservation projects in the area with support from Fundación Natura.
THREATS: agricultural encroachment from the slopes below.

10. San Miguel — Azuay

COORDINATES: 2°48'S, 79°30'W.
ALTITUDE: 900-1,500 m.
DESCRIPTION: relatively undisturbed tropical moist forest.
FOREST EXTENT: Large tracts remain (5,000 ha?), but precise forest extent not known.
SURVEY DATES: 8-14 Jan 1992 (MW); 14 observer days.
ORNITHOLOGICAL DATA: 99 species recorded, including two threatened species (*Onychorhynchus occidentalis* and *Lathrotriccus griseipectus*), and three near-threatened

cont.

San Miguel (cont.)

species (*Harpyhaliaetus solitarius, Aratinga erythrogenys* and *Brotogeris pyrrhopterus).*
PROTECTED STATUS: none.
THREATS: agricultural encroachment and selective logging.
REFERENCES: M. Whittingham *in litt.* (1992).

11. Uzhcurrumi — El Oro

COORDINATES: 3°21'S, 79°33'W.
ALTITUDE: 320-1,500 m.
DESCRIPTION: mostly degraded agricultural land, with some tiny patches of Semi-evergreen Intermontane Forest and Scrub along water courses.
FOREST EXTENT: no patches larger than 5 ha.
SURVEY DATES: 15 (NK and MK) and 22 Feb 1991 (BJB *et al.*); 6 observer days.
ORNITHOLOGICAL DATA: 52 species recorded, including three threatened species (*Leucopternis occidentalis, Pyrrhura orcesi* and *Attila torridus*) and three near-threatened species (*Aratinga erythrogenys, Brotogeris pyrrhopterus* and *Ampelioides tschudii*).
PROTECTED STATUS: none.
THREATS: complete clearance of the tiny forest patches that remain.
REFERENCES: Krabbe (1991), Best (1992).

12. Arenillas Military Reserve — El Oro

COORDINATES: 3°33'S, 80°03'W.
ALTITUDE: sea-level to 300 m.
DESCRIPTION: deciduous forest and mangroves.
FOREST EXTENT: not known; reserve extends to 20,000 ha.
SURVEY DATES: 13-14 July 1991 (TAP); 2 observer days.
ORNITHOLOGICAL DATA: 123 species recorded, including two threatened species (*Ortalis erythroptera* and *Synallaxis tithys*) and three near-threatened species (*Crypturellus transfasciatus, Aratinga erythrogenys* and *Brotogeris pyrrhopterus*).
PROTECTED STATUS: managed by the Ecuadorian army. From 1993 the Fundación Ecuatoriano para el Desarollo e Investigación del Medio Ambiente (FEDIMA) has been working with the army to prepare a management plan and seek financial support for the area.
THREATS: selective logging.
REFERENCES: Parker and Carr (1992), R. Phillips *in litt.* (1994).

13. BUENAVENTURA — El Oro

(9.5 km West of Piñas, Piñas)

COORDINATES: 3°40′S, 79°44′W.
ALTITUDE: 900-1,000 m.
DESCRIPTION: patches of Very Humid Premontane Cloud-forest c.9 km W of Piñas.
FOREST EXTENT: c.100 ha in total.
SURVEY DATES: Aug 1980 (ANSP), 11 June-6 July 1985 (ANSP), Aug 1988 (ANSP), Jan 1989 (ANSP), April 1989 (ANSP), Aug 1989 (ANSP), 25-26 Sept 1990 (NK), 23 Feb-6 Mar 1991 (BJB *et al.*), 14-16 April 1990, 7-9 Sept 1991 (RSRW and JAT), 15 Nov 1991 (NK and TSS), 8 Dec 1991 (EK and NK). Also many recent visits by bird tour groups.
ORNITHOLOGICAL DATA: at least 200 species recorded, including nine threatened species (*Leucopternis occidentalis, Ortalis erythroptera, Penelope barbata, Leptotila ochraceiventris, Pyrrhura orcesi, Cephalopterus penduliger, Onychorhynchus occidentalis, Lathrotriccus griseipectus* and *Attila torridus*) and three near-threatened species (*Harpyhaliaetus solitarius, Aratinga erythrogenys* and *Ampelioides tschudii*).
PROTECTED STATUS: none.
THREATS: selective logging, forest destruction for grazing land.
REFERENCES: Robbins and Ridgely (1990), Krabbe (1991), Best (1992), N. Krabbe *in litt.* (1992), M. B. Robbins *in litt.* (1992), Williams and Tobias (1994).

14. VICENTINO — Loja

COORDINATES: 3°57′S, 79°57′W.
ALTITUDE: 900-1,450 m.
DESCRIPTION: patches of Semi-evergreen Lowland and Premontane Tall Forest surrounded by crop-land and cattle-pastures to the NE and SW of Vicentino village.
FOREST EXTENT: forest patches do not exceed 20 ha.
SURVEY DATES: 8 and 14-18 Feb 1991 (BJB *et al.*); 9.5 observer days.
ORNITHOLOGICAL DATA: 87 species recorded, including four threatened (*Leucopternis occidentalis, Ortalis erythroptera, Leptotila ochraceiventris* and *Myrmeciza griseiceps*) and two near-threatened species (*Aratinga erythrogenys* and *Brotogeris pyrrhopterus*).
PROTECTED STATUS: none.
THREATS: further destruction of forest patches for agricultural land.
REFERENCES: Best (1992).

15. ALAMOR — Loja

COORDINATES: 4°02′S, 80°02′W.
ALTITUDE: 1,100-1,450 m.
DESCRIPTION: patches of Semi-evergreen Lowland and Premontane Tall Forest NE and W of Alamor, and agricultural land with scattered trees and hedges.
FOREST EXTENT: separate forest patches do not exceed 50 ha.

cont.

Alamor (cont.)

SURVEY DATES: 13-14 and 18-19 Feb 1991 (BJB *et al.*), 17, 25-31 Aug and 10 Sept 1991 (RSRW and JAT); 20.5 observer days (also early 20th century surveys by AMNH: Chapman 1926).
ORNITHOLOGICAL DATA: eight threatened species (*Leucopternis occidentalis, Leptotila ochraceiventris, Acesrura bombus, Synallaxis tithys, Syndactyla ruficollis, Hylocryptus erythrocephalus, Myrmeciza griseiceps* and *Attila torridus*) and two near-threatened species (*Ortalis erythroptera* and *Aratinga erythrogenys*).
PROTECTED STATUS: none.
THREATS: further destruction of forest patches for agricultural land.
REFERENCES: Best (1992), Williams and Tobias (1994).

16. TIERRA COLORADA — Loja

COORDINATES: 4°02'S, 79°57'W.
ALTITUDE: 1,400-1,850 m.
DESCRIPTION: a patch of Humid Lower Montane Cloud-forest at a valley head, surrounded by agricultural land.
FOREST EXTENT: 70 ha.
SURVEY DATES: 9-19 Feb 1991 (NK and FL; BJB *et al.*) 36 observer days.
ORNITHOLOGICAL DATA: 123 species recorded, including four threatened species (*Leucopternis occidentalis, Ortalis erythroptera, Syndactyla ruficollis* and *Attila torridus*) and three near-threatened species (*Tigrisoma fasciatum, Aratinga erythrogenys* and *Ampelioides tschudii*).
PROTECTED STATUS: none.
THREATS: complete clearance for agriculture.
REFERENCES: Best (1992).

17. CATACOCHA — Loja

COORDINATES: 4°03'S, 79°40'W.
ALTITUDE: 1,400-1,850 m.
DESCRIPTION: a small patch of Semi-evergreen Lower Montane Cloud-forest on a steep cliff-slope, with agricultural land below.
FOREST EXTENT: 40 ha.
SURVEY DATES: 4-5 Mar and 7-8 Mar 1991 (MC *et al.*), 2-3 Apr 1992 (ANSP).
ORNITHOLOGICAL DATA: four threatened species (*Ortalis erythroptera, Leptotila ochraceiventris, Syndactyla ruficollis* and *Hylocryptus erythrocephalus*) and two near-threatened species (*Aratinga erythrogenys* and *Brotogeris pyrrhopterus*).
PROTECTED STATUS: none.
THREATS: agricultural encroachment, logging.
REFERENCES: Best (1992), M. B. Robbins *in litt.* (1992).

18. Celica **Loja**

COORDINATES: 4°05'S, 79°57'W.
ALTITUDE: 1,600-2,800 m.
DESCRIPTION: several patches of Humid Montane Cloud-forest to the E, W and N of the town.
FOREST EXTENT: largest patches do not exceed 50 ha.
SURVEY DATES: 18 Nov 1988 (PC), 19-20 Feb 1989 (PC), 1 Mar, 25-27 Mar and 28 May-3 June 1989 (HB *et al.*), 11-12 Apr and 1 May 1989 (PC), 28 May-1 June 1989 (HB *et al.*), Aug 1989, 19-20 Sept 1990 (PC), 17,26 and 29-30 Mar 1991 (WFVZ), 6-8, 14 and 20 Feb 1991 (BJB *et al.*), 16 Aug, 10-11 Sept 1991 (RSRW and JAT). Also recent surveys by NK, ANSP and bird tour groups, and early 20th century surveys by AMNH: Chapman (1926).
ORNITHOLOGICAL DATA: eight threatened species (*Leucopternis occidentalis, Ortalis erythroptera, Leptotila ochraceiventris, Acestrura bombus, Syndactyla ruficollis, Hylocryptus erythrocephalus, Myrmeciza griseiceps* and *Attila torridus*) and one near-threatened species (*Aratinga erythrogenys*).
PROTECTED STATUS: none, however Fundación Arcolris, a local NGO, is developing a conservation proposal for the area with BirdLife International to seek funding for formal protection.
THREATS: further clearance for agricultural land.
REFERENCES: R. S. Ridgely *in litt.* to ICBP (1989), Best (1992), Bloch *et al.* (1991), P. Coopmans *in litt.* (1991), Kiff (1991), Best (1992), R. Phillips *in litt.* (1994), Williams and Tobias (1994).

19. El Empalme **Loja**

COORDINATES: 4°08'S, 79°49'W.
ALTITUDE: 700-900 m.
DESCRIPTION: Deciduous *Ceiba trichistandra* Forest, mostly with a sparse understorey. Large tracts, but poor quality.
FOREST EXTENT: only 30 ha of good quality forest along streams/ravines.
SURVEY DATES: 16-25 Aug 1989 (ANSP), 7 and 17 Feb 1991 (BJB *et al.*), 24 Aug 1991 (JAT & RSRW).
ORNITHOLOGICAL DATA: two threatened species (*Leptotila ochraceiventris* and *Hylocryptus erythrocephalus*) and two near-threatened species (*Aratinga erythrogenys* and *Brotogeris pyrrhopterus*).
PROTECTED STATUS: none.
THREATS: disturbance of forest understorey in the last areas where it is currently intact.
REFERENCES: M. B. Robbins *in litt.* (1991), Best (1992), Williams and Tobias (1994).

20. Sabanilla-Zapotillo — Loja

COORDINATES: 4°14'S, 80°11'W.
ALTITUDE: 500-550 m.
DESCRIPTION: *Acacia* forest and scrub, surrounded by agricultural land. Surveyed during an El Niño year when the vegetation was unusually lush.
FOREST EXTENT: extensive tracts of deciduous forest occur in the region, but most of the forest has a very open, degraded understorey.
SURVEY DATES: 8-9 Apr 1992, and April 1993 (ANSP).
ORNITHOLOGICAL DATA: three threatened (*Synallaxis tithys, Hylocryptus erythrocephalus* and *Carduelis siemiradzkii*) and two near-threatened species (*Crypturellus transfasciatus* and *Aratinga erythrogenys*).
PROTECTED STATUS: none.
THREATS: further degradation and fragmentation.
REFERENCES: M. B. Robbins *in litt.* (1992).

21. Sozoranga — Loja

COORDINATES: 4°21'S, 79°47'W.
ALTITUDE: 1,600-2,615 m.
DESCRIPTION: patches of Semi-evergreen Lower Montane Cloud-forest in ravines, surrounded by agricultural land and scrub.
FOREST EXTENT: patches not exceeding 30 ha.
SURVEY DATES: 10-12 June 1989 (HB *et al.*), 8-20 Aug and 8-21 Sept 1989 (BJB *et al.**), 19-23 Jul 1990 (RSRW and JAT), 30 Jan-1 Feb, 5-6 Mar and 9-12 Mar 1991 (BJB *et al.*); 145 observer days. Also Dec 1994 surveys by E. Barnes and G. Engblom.
ORNITHOLOGICAL DATA: 94 species recorded, including five threatened species (*Ortalis erythroptera, Leptotila ochraceiventris, Syndactyla ruficollis, Hylocryptus erythrocephalus,* and *Lathrotriccus griseipectus*), and three near-threatened species (*Harpyhaliaetus solitarius, Aratinga erythrogenys* and *Brotogeris pyrrhopterus*).
PROTECTED STATUS: one small forest patch is apparently being protected for watershed purposes, and Fundación ArcoIris, a local NGO is developing a conservation proposal for the area with BirdLife International to seek funding for formal protection.
THREATS: agricultural expansion, logging, soil erosion.
REFERENCES: Best and Clarke (1991), Bloch *et al.* (1991), Best (1992), R. Phillips *in litt.* (1994), Williams and Tobias (1994).

22. Utuana — Loja

COORDINATES: 4°22′S, 79°43′W.
ALTITUDE: 2,500 m.
DESCRIPTION: an area of Humid Montane Cloud-forest, surrounded by agricultural land.
FOREST EXTENT: 100 ha.
SURVEY DATES: 13-14 and 21-25 Sept 1989 (BJB *et al.**), 5-6 Feb 1991 (BJB *et al.*), 23 July 1991 (PC); 17 observer days.
ORNITHOLOGICAL DATA: At least 50 species recorded, including two threatened species (*Syndactyla ruficollis* and *Myrmeciza griseiceps*) and one near-threatened species (*Aratinga erythrogenys*).
PROTECTED STATUS: none.
THREATS: understorey removal, agricultural expansion.
REFERENCES: Best and Clarke (1991), P. Coopmans *in litt.* (1991), Best (1992).

23. Tambo Negro — Loja

COORDINATES: 4°24′S, 79°51′W.
ALTITUDE: 600-1,000 m.
DESCRIPTION: a large tract of Deciduous *Ceiba trichistandra* Forest on a ridge, extending into Peru.
FOREST EXTENT: 2,500 ha.
SURVEY DATES: 24 Aug-7 Sept and 26-30 Sept 1989 (BJB *et al.**), 26 Jan-7 Feb and 6-9 Mar 1991 (BJB *et al.*); 113 observer days. Also Dec 1994 surveys by C. Balchin, E. Barnes and G. Engblom.
ORNITHOLOGICAL DATA: 113 species recorded, including seven threatened species (*Ortalis erythroptera, Leptotila ochraceiventris, Synallaxis tithys, Syndactyla ruficollis, Hylocryptus erythrocephalus, Myrmeciza griseiceps* and *Lathrotriccus griseipectus*), and three near-threatened species (*Crypturellus transfasciatus, Aratinga erythrogenys* and *Brotogeris pyrrhopterus*).
PROTECTED STATUS: none, but the area is being managed by locals for hunting and cattle grazing.
THREATS: agricultural expansion and logging.
REFERENCES: Best and Clarke (1991), Best (1992), R. Phillips *in litt.* (1994).

24. Angashcola — Loja

COORDINATES: 4°34′S, 79°22′W.
ALTITUDE: 2,500-3,100 m.
DESCRIPTION: Humid Montane Cloud-forest.
FOREST EXTENT: 300-400 ha.
SURVEY DATES: 1-17 and 30 Aug-3 Sept 1990, 22-28 Jul 1991 (RSRW & JAT); 71 observer days.

cont.

Angashcola (cont.)

ORNITHOLOGICAL DATA: four threatened species (*Penelope barbata, Leptosittaca branickii, Metallura odomae* and *Syndactyla ruficollis*), and two near-threatened species (*Aratinga erythrogenys* and *Saltator cinctus*).
PROTECTED STATUS: none.
THREATS: logging and grazing of the understorey, hunting.
REFERENCES: Williams and Tobias (1994).

25. Tumbes National Forest — Tumbes

COORDINATES: 3°49'S, 80°17'W.
ALTITUDE: 400-750 m.
DESCRIPTION: the northern part of the North-West Peru Biosphere Reserve, forming the largest continuous tract of forest in the Tumbesian region. Forest types include Deciduous *Ceiba trichistandra* Forest, Semi-evergreen *Ceiba pentandra* Forest and Semi-evergreen Lowland and Premontane Tall Forest.
FOREST EXTENT: 75,102 ha in the reserve.
SURVEY DATES: 14 Jun-5 July 1979 (DAW *et al.*), 25 Feb-3 Mar 1986 (MK), 23-27 July 1988 (TAP and TSS).
ORNITHOLOGICAL DATA: 163 species recorded, including 11 threatened species (*Leucopternis occidentalis, Ortalis erythroptera, Leptotila ochraceiventris, Synallaxis tithys, Syndactyla ruficollis, Hylocryptus erythrocephalus, Myrmeciza griseiceps, Onychorhynchus occidentalis, Lathrotriccus griseipectus, Attila torridus* and *Carduelis siemiradzkii*) and four near-threatened species (*Crypturellus transfasciatus, Aratinga erythrogenys, Brotogeris pyrrhopterus* and *Pachyramphus spodiurus*).
PROTECTED STATUS: Peruvian government designated National Forest and Biosphere Reserve.
THREATS: settlement, logging, hunting.
REFERENCES: Wiedenfeld *et al.* (1985), Pulido (1991), M. Kessler *in litt.* (1992), Parker *et al.* (1995).

26. Cerros de Amotape National Park — Tumbes

COORDINATES: 4°28'S, 80°40'W.
ALTITUDE: 200-600 m.
DESCRIPTION: part of the North-West Peru Biosphere Reserve.
FOREST EXTENT: not known; Park extends to 91,300 ha.
SURVEY DATES: 11-18 Nov 1972 (LSUMZ).
ORNITHOLOGICAL DATA: two threatened species (*Syndactyla ruficollis* and *Lathrotriccus griseipectus*) and one near-threatened species (*Aratinga erythrogenys*) have been found but

cont.

Cerros de Amotape National Park (cont.)

no detailed surveys have been conducted.
PROTECTED STATUS: Peruvian government designated National Park.
THREATS: settlement. logging, hunting.
REFERENCES: Schulenberg and Parker (1981).

27. AYABACA — Piura

COORDINATES: 4°36′S, 79°44′W.
ALTITUDE: 2,625 m.
DESCRIPTION: Humid Montane Cloud-forest with cattle pastures.
FOREST EXTENT: a total of c. 100 ha in several patches.
SURVEY DATES: 22-26 Sept 1989 (CTC and MC); 7 observer days.
ORNITHOLOGICAL DATA: 44 species recorded including four threatened species (*Penelope barbata, Leptotila ochraceiventris, Syndactyla ruficollis* and *Myrmeciza griseiceps*).
PROTECTED STATUS: none.
THREATS: logging, agricultural encroachment.
REFERENCES: Best and Clarke (1991).

28. CANCHAQUE-HUANCABAMBA — Piura

(Inclues Cruz Blanca)

COORDINATES: 5°23′S, 79°37′W.
ALTITUDE: 1,700-3,500 m.
DESCRIPTION: a large area of cloud-forest and scrub, notable for its rare Andean species, but also supporting three important Tumbesian endemics.
FOREST EXTENT: not known.
SURVEY DATES: 25 Nov-10 Dec 1974, 19-25 Aug 1975, 11-20 Oct 1977, 7-12 July 1978, 10 June-25 July 1980 (LSUMZ)(also early 20th century surveys by AMNH and recent visits by birdwatchers); 380+ observer days.
ORNITHOLOGICAL DATA: six threatened species (*Penelope barbata, Hapalopsittaca pyrrhops, Syndactyla ruficollis, Hylocryptus erythrocephalus, Myrmeciza griseiceps* and *Buthraupis wetmorei*) and eight near-threatened species (*Harpyhaliaetus solitarius, Gallinago imperialis, Aegolius harrisii, Aratinga erythrogenys, Metallura odomae, Grallaricula peruviana, Myiophobus lintoni* and *Incaspiza ortizi*).
PROTECTED STATUS: none.
THREATS: cattle trampling and grazing of understorey, forest clearance for agriculture.
REFERENCES: Parker *et al.* (1985).

29. Olmos-Bagua **Lambayeque**

COORDINATES: 5°50'S, 80°17'W.
ALTITUDE: 500 m.
DESCRIPTION: semi-evergreen forest fragments in a ravine.
FOREST EXTENT: very small (a few hectares only).
SURVEY DATES: several LSUMZ surveys between 1964 and 1979; 13 Feb 1986 (MK and TM).
ORNITHOLOGICAL DATA: three threatened species (*Leptotila ochraceiventris, Syndactyla ruficollis* and *Hylocryptus erythrocephalus*) and one near-threatened species (*Crypturellus transfasciatus*).
PROTECTED STATUS: none.
THREATS: deforestation.
REFERENCES: Schulenberg and Parker (1981), M. Kessler *in litt.* (1993).

30. Quebrada Caballito **Piura**

(río Tocto valley)

COORDINATES: 5°53'S, 80°19'W.
ALTITUDE: 500-1,000 m.
DESCRIPTION: mostly deciduous (thorn) forest.
FOREST EXTENT: at least 100 ha.
SURVEY DATES: 9-12 Feb 1986 (MK & TM); 6 observer days.
ORNITHOLOGICAL DATA: two threatened species (*Penelope albipennis* and *Hylocryptus erythrocephalus*).
PROTECTED STATUS: private reserve.
THREATS: further degradation.
REFERENCES: M. Kessler *in litt.* (1992).
Note: several other small valleys in the range of *Penelope albipennis* may also support *Hylocryptus erythrocephalus* and *Aratinga erythrogenys.*

A summary table listing these important ornithological sites and their priority species complements appears below (Figure 87).

Figure 87. Important ornithological sites in the Tumbesian region and their species complements. Each locality listed in the directory is included.

Sites	Species																				
	PBT	GBH	RHC	WWG	.OBD	RMP	EOP	GCP	EW	BHS	RNFG	HHFG	GHA	PRF	GBF	PCT	TT	OA	SB	PP	PHBF	SS
1. Cerro Mutiles		■	■																			
2. Cabeceras de Blisa		■	■															■				
3. Río Palenque		■	■			■								■	■			■	■			
4. Jauneche	■		■		■	■								■	■			■				
5. Hacienda Pacaritambo		■						■						■				■	■			
6. Machalilla N.P.	■	■	■		■	■			■	■		■		■	■			■				■
7. Cerro Blanco	■	■	■		■	■		■		■		■		■	■							■
8. Manglares-Churute E.R.		■	■		■	■		■						■	■							
9. Manta Real		■	■			■	■	■						■	■			■	■			
10. San Miguel						■		■						■	■							
11. Uzhcurrumi		■				■	■	■										■				
12. Arenillas M. R.	■		■			■		■		■												
13. Buenaventura		■	■		■	■	■							■	■			■				
14. Vicentino		■	■		■	■		■					■									
15. Alamor		■	■		■	■				■	■	■	■					■				
16. Tierra Colorada		■	■			■					■							■				
17. Catacocha			■		■	■		■			■	■										
18. Celica		■	■		■	■					■	■	■					■				
19. El Empalme					■	■		■				■										
20. Sabanilla-Zapotillo	■					■				■		■										■
21. Sozoranga			■		■	■		■			■	■			■							
22. Utuana						■					■		■									
23. Tambo Negro	■		■		■	■		■		■	■	■	■		■							
24. Angashcola						■					■											
25. Tumbes National Forest	■	■	■		■	■		■		■	■	■	■	■	■			■	■			■
26. Cerros de Amot. N. P.						■					■				■							
27. Ayabaca					■						■		■									
28. Canchaque-Huanc.						■					■		■									
29. Olmos-Bagua	■				■						■	■										
30. Quebrada Caballito				■								■										

Sources: Brossett (1964), Vuilleumier (1978), Leck (1979), Leck *et al.* (1980), Schulenberg and Parker (1981), Wiedenfeld *et al.* (1985), MacBryde (1987), P. J. Greenfield *in litt.* to ICBP (1989), R. S. Ridgely *in litt.* to ICBP (1989), Best and Clarke (1991), Bloch *et al.* (1991), P. Coopmans *in litt.* (1991), Kiff (1991), Krabbe (1991), Ridgely (1991a, 1991b), R. S. Ridgely *in litt.* (1991), Best (1992), P. Coopmans *in litt.* (1992), M. Kessler *in litt.* (1992), N. Krabbe *in litt.* (1992), Parker and Carr (1992), M. B. Robbins *in litt.* (1992), M. Whittingham *in litt.* (1992), K. S. Berg *in litt.* (1993), M. Kessler *in litt.* (1993), Williams and Tobias (1994), Parker *et al.* (1995).

see Table 3 for explanation of abbreviated species names

CONSERVATION RECOMMENDATIONS

INTRODUCTION

THE PRECEDING chapters have demonstrated the importance of the Tumbesian region in terms of the high levels of endemism of its flora and avifauna, and the unusually wide diversity of vegetation types it supports. However, human pressures have brought many of the endemic species to the verge of extinction and very little original forest remains. This calls for urgent conservation action.

This chapter aims to use all available biological data yet the minimum of resources, to set out the action necessary to maintain maximum biological diversity within the Tumbesian Centre of Endemism. Two different approaches are combined. Because the region's forests have been so severely degraded there is a danger that within the next decade whole habitats will be lost, along with the species which are restricted to them. This requires an immediate conservation effort: the designation of areas worthy of particular protection as reserves. Some of these areas have already been given reserve status; others will need to be set up as new reserves. These are areas which support populations of threatened species, vegetation types or ecosystems. Secondly, as the protection of individual areas is doomed to fail unless the local people and the environmental situation of the whole region are taken into account, a regional environmental programme must also be implemented. We believe this latter approach is the only way to achieve the long-term conservation of the biodiversity Tumbesian region, but it will necessarily take time to implement; in the meantime the few valuable tracts of remaining habitat may be lost. For this reason protected areas are also a vital component.

As there are many gaps in our knowledge of habitat and species distributions in the Tumbesian region, the recommendations presented are not definitive, rather they are based on what information is available now. We simply cannot afford to wait until our knowledge of the region is complete. Included in this chapter is a section detailing the most important research priorities which remain, themselves an integral part of any conservation measures.

In order to protect the maximum numbers of species in the Tumbesian region it is important that all the characteristic vegetation types are safeguarded. In the short term an effective way of achieving this is by creating reserves. These conservation recommendations aim to protect at least two representative samples of all the important vegetation types inside reserves. Two reserves are preferable over one because having only one for a particular vegetation type increases the risk of destruction through freak events (e.g. fire) or intense human pressure, or the genetic deterioration of the site (Wilcox 1982). It also increases the genetic variability of the ecosystem; it should however be seen as the minimum conservation action necessary. Because so little is known about plant and animal distributions in the region, it is not possible to know whether the recommendations meet the needs of all the endemic taxa (a discussion for birds appears below), but by focusing on discrete habitats a range of species should be protected. Furthermore it is flexible: if 'new' habitat patches or new populations of endemic species are discovered in future, these can easily be incorporated.

WHICH HABITATS MUST BE PROTECTED?

The 'Vegetation' chapter showed that some of the vegetation types of the Tumbesian region have higher conservation importance than others due to their uniqueness, degree of endemicity and species richness. Two priority groups for conservation were identified (subsequently collectively called "the priority vegetation types"; Box 6), each containing four vegetation types. Habitats in group one have higher conservation priority than those in group two.

Table 11 shows that four (50%) of the priority vegetation types (all group one) are already protected within at least two existing reserves. These existing protected areas are clearly extremely important conservation sites and have a significant roles to play in preserving biodiversity in the Tumbesian region. The next section details the action necessary to safeguard these crucial sites.

ACTION REQUIRED IN THE EXISTING TUMBESIAN RESERVES

All existing reserves in the Tumbesian region are important from the point of view of habitat protection so they must be securely protected and effectively managed. Each has its own combination of specific threats, the most prevalent being the slow degradation of the forest and its understorey by agricultural encroachment from neighbouring lands and roaming cattle and feral animals, which destroy the understorey and suppress tree regeneration. The managers in charge of each protected area should be informed of the reserve's vital role in preserving the

Box 6. Priority habitat types of the Tumbesian region.

Group 1: Conservation essential

Mainly Deciduous Tropical Thorn-forest and *Acacia* Thorn-Forest*
Mainly Deciduous *Ceiba trichistandra* Forest*
Semi-evergreen *Ceiba pentandra* Forest*
Semi-evergreen Lowland and Premontane Tall Forest*

Group 2: Conservation very important

Humid to Very Humid Premontane Cloud-forest
Deciduous to Semi-evergreen Lower Montane Cloud-forest
Humid Coastal-hill Cloud-forest
Deciduous to Semi-evergreen Intermontane Scrub, Thorn- forest and Forest

Table 10. The vegetation types of the existing Tumbesian protected areas.

Vegetation type	Protected areas							
	1	2	3	4	5	6	7	8
Group A								
1. Mainly Deciduous Tropical Thorn-forest and *Acacia*Thorn-Forest	*	?	-	-	-	-	*	2+
2. Mainly Deciduous *Ceiba trichistandra* Forest	*	*	*	*	-	*	-	5
3. Semi-evergreen *Ceiba pentandra* Forest and Semi-evergreen Lowland and Premontane Tall Forest	-	?	*	*	-	-	?	2+
Group B								
4. Humid to Very Humid Premontane Cloud-forest	-	-	-	-	-	-	-	0
5. Deciduous to Semi-evergreen Lower Montane Cloud-forest and Forest	-	-	-	-	-	-	-	0
6. Humid Coastal-hill Cloud-forest	-	-	-	*	-	-	-	1
7. Deciduous to Semi-evergreen Intermontane Scrub, Thorn-forest and forest								0
Group C								
8. Semi-desert	-	-	-	-	-	-	-	0
9. Moist Lowland Forest	-	-	-	-	*	-	-	1
10. Humid to Very Humid Lower Montane Cloud-forest	-	-	-	-	-	-	-	0
11. Humid to Very Humid Montane Cloud-forest	-	-	-	-	-	-	-	0
Number of vegetation types per reserve	2	1+	2	3	1	1	1+	1+

Protected areas: 1. El Angolo Hunting Reserve, 2. Cerros de Amotape National Park, 3. Tumbes National Forest (these three areas together form the North-West Peru Biosphere Reserve which embraces three vegetation types), 4. Machalilla National Park, 5. Jauneche Reserve, 6. Cerro Blanco Reserve, 7. Quebrada Negrohuasi, 8. Number of reserves in which the vegetation type occurs.

biodiversity of the Tumbesian Centre of Endemism and preferably be given outside technical and financial support in order to preserve their integrity. Efforts must be made to reduce the pressure on the reserves. This could be achieved by planting quick-growing trees outside reserve boundaries to reduce wood-extraction pressure within the reserve itself, and by creating buffer zones around each reserve in which only limited uses are allowed (depending on the reserve), as has been tried in the Machalilla National Park. Environmental education programmes in the local area of each reserve should aim to stress the local importance of forests as water-catchment areas and promote their value in preventing soil erosion and nutrient loss (see below under 'Other action'). It is important that local people see the reserves as having tangible value for them; if possible they should be employed as guards or guides for visiting ecologists (and tourists if ecotourism is feasible). Management objectives of existing reserves should be assessed and where appropriate technical and financial support should be secured to improve management. Specific threats to the existing protected areas are detailed below.

Machalilla National Park

Although one of the unique ecological areas of Ecuador, this reserve suffers many human pressures (Salazar and Huber 1982). Much dry forest in this reserve has been degraded and wildlife is threatened by colonization, logging, hunting, agriculture and animal grazing (MacBryde 1987, Dodson and Gentry 1991, Figueroa 1992). Semi-wild domestic animals roam everywhere in the park (mainly goats and mules; Salazar and Huber 1982), suppressing tree regeneration; these authors stress that they must be controlled by their owners or eradicated from the park. Park managers are attempting to limit settlement to certain areas which are then removed from the park area (Salazar and Huber 1982), but this has been complicated by confusion over the ownership of certain areas. Soil erosion continues to be a problem, causing the sedimentation of rivers. Reforestation of degraded parts of the park could be attempted. It is important to establish a research centre to monitor the status of the park, study its wildlife and conduct inventories.

North-West Peru Biosphere Reserve

This area is apparently only protected by its remoteness at present and is at risk from future settlement, logging and agriculture, since a road is being re-opened into the area after being destroyed by the 1982-1983 El Niño floods (Parker *et al.* 1989). The reserve is seemingly only guarded by military personnel who patrol the border with Ecuador. An important recent development is the setting up of a research centre in the southern part of the Biosphere Reserve in the El Angolo Hunting Reserve, by the National Agricultural University of La Molina. This is used primarily by foresters although some mammal studies have been undertaken there;

it is important to establish a research station similar to the one at La Molina in the Tumbes National Forest.

Both the Machalilla National Park and the North-West Peru Biosphere Reserve also suffer from understaffing and lack of equipment and funds. The highest conservation priority should be given to making these two areas secure. They are the two largest remaining areas of forest in the Tumbesian region; the North-West Peru Biosphere Reserve standing out in terms of its size and comparatively pristine forest. Outside funds should be injected into increasing the staffing levels of the parks as well as the resources available to the park guards and managers. An effort to secure the long-term future of these two areas is an essential foundation of any conservation measures for the Tumbesian region.

Manglares-Churute Ecological Reserve

The biological importance of this area has yet to be established. It covers some 35,000 ha, although the exact amount of forest is not known. The reserve is known to contain forest of a suitable type for threatened Tumbesian bird species (though perhaps rather limited in extent); four have been found to date: *Leucopternis occidentalis, Leptotila ochraceiventris, Onychorhynchus occidentalis* and *Lathrotriccus griseipectus* (R. S. Ridgely *in litt.* 1992), but there appear to have been very few specific ornithological surveys. K. S. Berg and N. Hilgert have carried out recent surveys there and these investigations should continue as the area may prove to have considerable ornithological and botanical value.

Other protected areas

In addition to the above comparatively large protected areas, there are five smaller, privately owned reserves in the Tumbesian region, four in Ecuador: Cerro Mutiles (Esmeraldas Province), Río Palenque (Pichincha Province), Jauneche (Los Ríos Province) and Cerro Blanco (Guayas Province); and one in Peru: Quebrada Negrohuasi (Lambayeque Department). These smaller protected areas are also threatened by agricultural encroachment, roaming cattle and illegal deforestation; their wildlife is also subject to hunting. The owners of these private reserves should be encouraged in their efforts to keep each secure and informed of the value of their forests.

Reforestation

As a more general recommendation, reforestation schemes could be attempted in the degraded parts of larger reserves (such as the Machalilla National Park), or in neighbouring areas of smaller reserves. Such attempts should, however, be guided by experience gained from dry forest regeneration elsewhere (e.g. Costa Rica) as the successful restoration of dry forest is particularly difficult. Janzen (1988) pointed out that many forest fragments are "decomposing", i.e. many of the component plant species cannot regenerate and will eventually become locally extinct. This is caused by human impact (especially domestic animal grazing) and by the small size of the habitat fragments. In the case of trees, species may take hundreds of years to disappear, so the process may not be obvious. When a degraded or completely cleared forest is allowed to grow back, its species composition will depend on which species will colonize the area first. If there are individual seed trees left, these will dominate the regeneration. In the case of large (over 1 km^2) clearings, wind-dispersed trees will dominate the newly growing forest. Animal-dispersed trees, especially those dispersed by forest-dwelling vertebrates, will be under-represented in large clearings, as the dispersing animals will tend to stay out of them.

Once a forest has been established, the dominant tree species will tend to continue as such, and it will take many tree generations before animal-dispersed trees will make a significant contribution to the species composition of the new forest, if in fact they ever do. It is possible that the species composition and dominance of tropical forests is based largely on stochastic events (e.g. Connell 1978, Hubbell and Foster 1986, Gentry 1988); in other words there is no fixed climax species composition, but if a forest by chance has a certain composition, it will tend to stay as it is or change by chance. The animal-dispersed trees are of great importance for the survival of those animals which feed on them, e.g. Terborgh (1986) found that in an Amazonian forest area a group of only 12 palm and fig species (out of 2,000 plant species) maintains almost all large frugivores for about three months of the year.

Therefore if a forest patch is allowed to regenerate by itself it will end up with a different tree composition from one that is managed, e.g. by the introduction of particular tree species (such as fruiting trees, or valuable timber trees which are likely to be under-represented in all remaining Tumbesian forests). Janzen (1988) concluded that different forest patches should be managed in different ways in order to create a diverse mosaic of different forest types. It is encouraging that natural invasion of forest trees into cleared pastures seems to proceed at a much faster rate in dry forests than in wet forests (Janzen 1988).

This is just one example which shows that reserves not only have to be protected, but also must be effectively managed to ensure the survival of the species and communities that are to be preserved. Much baseline research is needed here (see research priorities). Such problems encountered in safeguarding the existing protected areas of the Tumbesian region effectively clearly emphasize that once established, additional funds will be required to ensure that any new reserves are kept secure.

THE NEED FOR NEW TUMBESIAN RESERVES

Although four priority vegetation types are already 'protected' inside at least two existing reserves, this leaves a further three without protection, while the final one is protected in only one reserve (Table 10). Table 11 shows a priority ranking of areas important for habitat conservation in the Tumbesian region. In addition to the existing protected areas (listed in priority order based on the habitat type(s) they support) five new Tumbesian reserves are proposed in priority order. Priorities are set consecutively, i.e. once a vegetation type is represented twice an area with a different vegetation type is preferred. The proposed reserves are all in Ecuador and Box 7 lists further information on them.

Box 7. Proposed new Tumbesian reserves based on habitat priorities
Reserves are listed in rank order.

1. Manta Real (2°34'S 79°21'W), Azuay Province.
For Humid to Very Humid Premontane Cloud-forest.

2. Hacienda Quesada (3°20'S 79°18'W), Azuay Province.
For Deciduous to Semi-evergreen Intermontane Scrub, Thorn-forest and Forest.

3. A reserve in the Sozoranga (4°21'S 79°47'W) or Catacocha (4°03'S 79°40'W) areas, Loja Province.
For Deciduous to Semi-evergreen Lower Montane Cloud-forest.

=4. Cabeceras de Bilsa (0°37N 79°51W), Esmeraldas Province.
For Humid Coastal-hill Cloud-forest. *The Jatun Sacha Bilsa area* (0°22'N, 79°45'W) *is an alternative site.*

=4. A reserve in the very humid forest of El Oro Province (site yet to be identified, potentially Buenaventura 3°40'S 79°44'W).
For Humid to Very Humid Premontane Cloud-forest.

Although we have prioritized action in specific areas it is important to emphasize that every single remaining patch of forest in the Tumbesian region is valuable as a refuge for plants and animals and in protecting water supplies. Each local community has a role to play, and a benefit to receive, in maintaining these forest fragments.

In addition to protecting habitats as a basis for conservation in the Tumbesian region, there are sound reasons for doing so from a plant conservation stand-point. A fundamental difference between patterns of endemicity in plants and that in birds is that the plants can have more restricted ranges than birds. For example, while

Table 11. Priority sites for habitat conservation in the Tumbesian region.

Site	Vegetation types (see Table 11)	Degree of destruction	Score	Priority	New reserve priority	Already protected
NW Peru Biosphere Reserve	1, 2, 3	A, C	9	1	-	YES
Machalilla National Park	2, 3, 4	C	8	2	-	YES
Manta Real	5	A	2	3	1	-
Hac. Quesada	6	B	2	4	2	-
Sozoranga/Catacocha	7	D	2	5	3	-
Arenillas	1, 2	C	6	6	-	YES
Cabeceras de Bilsa	4	A	2	=7	=4	-
El Oro	5	C	2	=7	=4	-
Q. San Isidro/Q. Negrohuasi	1	C	3	=8	-	YES
Tambo Negro/Zapotillo	2	D	3	=8	-	-
Celica/Alamor	10	D	1	+	-	-
Cerro Pata de Pajaro	4	B	2	9	-	-
Cerro Blanco	2	D	3	10	-	YES
Jauneche	9	B	1	+	-	YES
Río Palenque	9	D	1	+	-	YES
Cerro Mutiles	9	C	1	+	-	YES

+'extremital' vegetation types: conservation best achieved outside the Tumbesian region.
Score: calculated using vegetation type in group A=3 points, B=2 points, C=1 points (see Table 10).
Forest disturbance: A= large, little disturbed forest areas, B= small, little disturbed forest areas, C= large, disturbed forest areas, D= small, disturbed forest areas. (Large = >5,000 ha).
Sites in italics are proposed new reserves

about 1,200 plant species are endemic to the Ecuadorian provinces of Loja, El Oro and Azuay, only two bird species are restricted to this region (*Pyrrhura orcesi* an as yet undescribed *Scytalopus* species, N. Krabbe *in litt.* 1992). Many plant species are restricted to only one vegetation type, with some occurring only in part of the range of particular vegetation types. As shown in the 'Vegetation' chapter, lowland vegetation types tend to support species which are distributed throughout the range of their habitat type, while species of the Andean slopes show more pronounced local endemism, requiring the protection of several examples of these vegetation types if the complete flora is to be saved. This means that from a plant conservation viewpoint, it is vital to ensure that the distinctive vegetation types of the region are all protected, with emphasis on those vegetation types with the most marked endemicity.

HOW WELL DO THE PROPOSED HABITAT CONSERVATION MEASURES PROTECT THE ENDEMIC AVIFAUNA OF THE TUMBESIAN REGION?

Having suggested the proposed action necessary to preserve the habitats of the Tumbesian region, we will now consider whether this is sufficient to protect the endemic avifauna of the region, the only other taxonomic group known well enough to carry out such a comparison. The 'Avifauna' chapter identified 22 Tumbesian bird species in need of conservation action, 16 threatened species and six near-threatened species. Collectively these 22 species will be referred to as the "priority Tumbesian bird species". The remaining Tumbesian species are not discussed here because they seem able to survive at many secondary forest and scrub sites in the Tumbesian region and we believe, therefore, that they do not need any specific conservation action to ensure their continued persistence.

It is important to emphasize that habitat destruction in the Tumbesian region is in such an advanced state that there are few opportunities for experimentation or debate on the criteria for selecting or siting new reserves. The number of actual forest patches of sufficient size is so low that it could be reasonably recommended that each is of extreme importance for the conservation of the avifauna. However, certain areas stand out. The approach taken is a modified version of the "network analysis" or "critical faunas analysis" of Ackery and Vane-Wright (1984); the fauna under consideration being the endemic birds of the Tumbesian centre.

Table 12 establishes whether the habitat conservation recommendations provide at least two geographically separate areas for each near-threatened and threatened bird species. As with the habitat conservation recommendations, such an approach represents the minimum number of reserves necessary. The table shows which threatened and near-threatened Tumbesian species occur in the existing protected areas and the sites designated for protection under the habitat conservation recommendations. Twelve species occur in at least two of the existing protected areas of the Tumbesian region and require no additional reserves. Potential sites for the conservation of the remaining ten species will now be discussed.

One species occurs in two of the areas designated for protection in the habitat conservation recommendations.

El Oro Parakeet *Pyrrhura orcesi*
This rare parakeet is endemic to Ecuador and is confined to a small area in the humid forests of Azuay and El Oro Province between 300 and 1,300 m, and currently occurs in no protected areas. There is little choice when it comes to suggesting protected areas as it is currently only known from a handful of localities, one of which (Uzhcurrumi) is largely deforested. The creation of reserves in the Piñas area and at a second site in Azuay Province are priorities for this species. Recent research has identified a large area of humid forest in Azuay which would be suitable for a reserve and probably supports *P. orcesi* (M. Whittingham *in litt.* 1992). Additionally, its most northerly locality, Manta Real in Azuay Province, is apparently earmarked for protection by the Ecuadorian conservation organizations Fundación Natura and CECIA (P. Greenfield *in litt.* to ICBP 1989). However, to date no progress has been made and the area continues to be cleared (R. Phillips verbally 1992). The habitat conservation recommendations state that two areas critical for the survival of *Pyrrhura orcesi* be set up as new reserves because of their habitat value: an area of very humid forest in El Oro Province (potentially at Buenaventura) and Manta Real in Azuay Province.

Five species occur in at least one existing reserve or in one proposed new reserve suggested in the habitat conservation recommendations. A fourth has been found on the very edge of the Machalilla National Park and probably also occurs within the park boundary.

White-winged Guan *Penelope albipennis*
This species was present in the coastal mangroves of north-western Peru in 1886 and 1887, but was not recorded afterwards and believed extinct until a small population was found in the dry wooded valleys of northern Peru in 1977 (de Macedo 1978). There is a possibility that the bird occurs in the arid part of south-western Ecuador: in 1980 White-tailed Jays *Cyanocorax mystacalis* were heard giving imitations of guan alarm calls in dry forest in coastal El Oro; these referred to either *P. albipennis* or *P. purpurascens,* although the latter species normally

Table 12. Priority sites for bird conservation in the Tumbesian region.
The distributions of the threatened and near-threatened Tumbesian endemics in 18 areas are given.

Sites	PBT	GBH	RHC	WWG	OBD	RMP	EOP	GCP	EW	BHS	RNFG	HHFG	GHA	PRF	GBF	PCT	TT	OA	SB	PP	PHBF	SS
	Species																					
Existing protected areas or proposed new reserves based on habitat conservation priorities																						
NW Peru Biosphere Reserve	X	X	X		X	X		X		X	X	X	X	X	X			X	X			X
Machalilla National Park	X	X	X		X	X			X	X		X		X	X			X				X
Hacienda Quesada																						
Manta Real		X	X			X	X	X						X	X			X	X			
El Oro Province humid forest		X	X		X	X	X							X	X			X				
Sozoranga/Catacocha			X		X	X		X			X	X	X		X							
Celica/Alamor		X	X		X	X		X		X	X	X	X					X				
Arenillas	X		X			X		X		X												
Cabeceras de Bilsa		X	X															X				
Cerro Mutiles		X	X																			
Río Palenque		X	X			X								X	X			X	X			
Jauneche	X		X		X	X								X	X			X				
Cerro Blanco	X	X	X		X	X		X		X		X		X	X							X
Manglares-Churute E. R.		X	X		X	X		X						X	X							
Quebrada Negrohuasi				X								X										
Species occurs in two existing reserves	**X**	**X**	**X**		**X**	**X**		**X**		**X**		**X**		**X**	**X**			**X**				**X**
Species occurs in two existing or proposed reserves	**X**	**X**	**X**		**X**	**X**	**X**	**X**	**X**	**X**	**X**	**X**		**X**	**X**			**X**	**X**			**X**
Additional sites of ornithological interest																						
Celica		X	X		X	X					X	X	X					X				
Sozoranga			X		X	X		X			X	X			X							
Tambo Negro	X		X		X	X		X		X	X	X	X		X							

Sources: Leck (1979), Leck *et al.* (1980), Schulenberg and Parker (1981), Wiedenfeld *et al.* (1985), P. J. Greenfield *in litt.* to ICBP (1989), R. S. Ridgely *in litt.* to ICBP (1989), Best and Clarke (1991), Bloch *et al.* (1991), P. Coopmans *in litt.* (1991), Kiff (1991), Krabbe (1991), Ridgely (1991a, 1991b), R. S. Ridgely *in litt.* (1991), Best (1992), P. Coopmans *in litt.* (1992), M. Kessler *in litt.* (1992), M. Whittingham *in litt.* (1992), N. Krabbe *in litt.* (1992), Parker and Carr (1992), M. B. Robbins *in litt.* (1992), K. S. Berg *in litt.* (1993), M. Kessler *in litt.* (1993), Williams and Tobias (1994) and Parker *et al.* (1995).
see Table 3 for explanation of abbreviated species names

occupies humid forest in Ecuador (R. S. Ridgely *in litt.* to ICBP 1992). There are also reports of an unidentified *Penelope* species, either *P. albipennis* or *P purpurascens*, in the Tumbes National Forest (Parker *et al.* 1995). The mangrove forests bordering the Pacific Ocean in south-western Ecuador and north-western Peru may also harbour the species (these have been rapidly cleared for the shrimp-farming industry).

Surveys in all these sites are an urgent priority, along with the establishment of new reserves and environmental education programmes in its only known stronghold in Peru. Although a small reserve has already been set up in Peru (at Quebrada Negrohuasi), protection apparently exists on paper only (Collar *et al.* 1992), and practical measures should be taken to protect the reserve effectively. At other sites, such as Quebrada San Isidro, money has been invested to protect the species. The main problem is that it occurs at very low densities in numerous small valleys in northern Peru, none of which support viable populations, so that any reserves created would not hold many individuals.

Esmeraldas Woodstar *Acestrura berlepschi*

This very rare hummingbird is endemic to Ecuador and has only been recorded from the coastal west-central sector. Conservation action should be directed to the Coastal Cordillera, where it was known from, and rediscovered in 1990 (in the Ayampe area of Manabí Province), on the edge of the Machalilla National Park (Collar *et al.* 1992). A second area where a population can be protected should be identified, with searches directed towards the Coastal Cordillera to the north of Ayampe and the interior of the Machalilla National Park which has so far yielded no records, but possesses suitable habitat. Apparent seasonal movements of the species complicate any conservation action (it was not found in July 1992 at Ayampe where it had been found in March 1990 and also seen in January 1991: R. S. Ridgely *in litt.* to ICBP 1992). At present it is not possible to recommend any future reserves for this hummingbird.

Rufous-necked Foliage-gleaner *Syndactyla ruficollis*

This scarce furnariid has been principally found in two areas of south-west Ecuador, both within Loja Province: the Celica-Alamor massif; and the Utuana-Tambo Negro region near Sozoranga. The bird is also known from mountainous northern Peru (e.g. Huancabamba: Parker *et al.* 1985; Ayabaca: Best and Clarke 1991), as well as in the Tumbes National Forest, where it is uncommon and occurs towards the lower end of its preferred altitudinal range. The creation of reserves in both the Celica-Alamor and Utuana-Sozoranga regions would be desirable as each area supports a healthy complement of priority species (13 and 10 priority species respectively). Only the latter area has been designated for protection in the habitat conservation recommendations. On ornithological grounds it is important that the Celica-Alamor region is also protected as it is noteworthy for the humid montane habitat favoured by *Syndactyla ruficollis* and also *Myrmeciza griseiceps* (see below). The lower part of the Sozoranga region (especially Tambo Negro) is also important for its large tract of *Ceiba* forest inhabited by *Leptotila ochraceiventris* and *Synallaxis tithys*. The latter area is cooperatively owned and is being protected (as a source of water and hunting) by local people who grow rice in the river valley below. Recent fieldwork near Alamor has revealed a potential area for protection to the north-east of that town (R. Phillips verbally 1992). Additionally, maximizing the range of forest types which are protected would seem prudent in light of evidence of seasonal use of several forest types by certain species.

Grey-headed Antbird *Myrmeciza griseiceps*
This poorly known antbird has a similar distribution to *Syndactyla ruficollis*, and like that species favours the understorey of dry and humid forest. The best location for a reserve for this species would be in the Celica-Alamor region. The size and breeding status of the Tambo Negro population of *Myrmeciza griseiceps* should be established.

Slaty Becard *Pachyramphus spodiurus*
Outside the Tumbes National Forest and the Río Palenque Reserve, the only recent records of *P. spodiurus* all come from Ecuador: at Manta Real (Azuay Province), near Arenillas (El Oro Province), and Puyango (Loja Province) (Collar *et al.* 1992). Of these only Manta Real is designated for protection in the habitat conservation recommendations. A detailed search must be made in adjacent regions and the sizes of the populations in these areas should be established. The species is known from the Río Palenque reserve in Pichincha Province, Ecuador (F. Ortiz-Crespo *in litt.* 1992) although the tiny size of that reserve means that the overall population cannot be very large. An urgent priority is to establish whether it occurs in the Machalilla National Park. The most suitable site for a new reserve may be Manta Real, which is apparently designated for protection by Fundación Natura and CECIA (see above) and has been highlighted in the habitat conservation recommendations. The bird occurs close to the Celica-Alamor massif and may occur at lower elevations on the massif itself; if so it could be protected by the proposed reserve in that region.

The remaining four species occur in no existing or proposed reserve.

Piura Chat-Tyrant *Ochthoeca piurae*, **Tumbes Tyrant** *Tumbezia salvini* and

Peruvian Plantcutter *Phytotoma raimondii*
These three species pose special problems as they are known from only a handful of localities in north-western Peru and are extremely poorly-known birds. The two tyrants inhabit scrubby regions and riparian thickets, yet have very local distributions in the dry Andean foothills (*O. piurae*) and coastal lowlands (*T. salvini*). The plantcutter occupies similar habitats in coastal north-western Peru. They occur in a region which has not been well surveyed by ornithologists, which may partly explain why they have been so rarely encountered. As there is still fairly extensive suitable habitat in their range further records may be expected. The first priority is to find out more about their habitat preferences and determine the threats to their habitats. Only then can more precise conservation action be taken, if indeed it is necessary.

Pale-headed Brush-Finch *Atlapetes pallidiceps*
This gravely threatened brush-finch is endemic to Ecuador, has only been recorded from a tiny area in the provinces of Azuay and Loja, and may already be extinct. It was not found in its old localities during several recent, brief searches (e.g. March 1990, B. Whitney; February 1991, N. Krabbe and M. Kessler; October 1991, J. Tobias and R. S. R. Williams; March 1992, M. Robbins, G. Rosenberg and F. Sornoza; late 1992, N. Krabbe). A thorough search must be made in all suitable habitat, especially at the onset of the rainy season (when it was last seen near Oña in 1965), and any populations found should be stringently protected until its habitat requirements are understood.

A summary of the protected areas needed for the these species appears below (Table 13).

Conclusion

The recommendations for habitat conservation appear to meet the requirements of the Tumbesian avifauna rather well: 16 (73%) of the threatened and near-threatened Tumbesian endemics would be protected in at least two of the areas already highlighted. For an additional six species (detailed above) it is currently not possible to recommend reserves since they have rather local distributions in habitats or areas where few other key species occur. Further research is required on these species and they are not included in the discussion which follows.

Six sites are required to provide two or more protected areas for each threatened and near-threatened Tumbesian bird (Box 8). Although the bird conservation recommendations constitute a species by species approach, they are provided only as a supplement to the habitat conservation measures. It is hoped that by ensuring that the distinctive habitat types of the region are safeguarded, a protected area

Table 13. Proposed new Tumbesian reserves based on bird conservation priorities.

Species	Site(s) proposed
Penelope albipennis	Further sites in north-west Peru
Pyrrhura orcesi	Piñas area, *Azuay Province site*
Acestrura berlepschi	None currently appropriate
Syndactyla ruficollis	Celica-Alamor, *Utuana-Tambo Negro*
Myrmeciza griseiceps	Celica-Alamor
Ochthoeca piurae	None currently appropriate
Tumbezia salvini	None currently appropriate
Pachyramphus spodiurus	*Manta Real*
Phytotoma raimondii	None currently appropriate
Atlapetes pallidiceps	None currently appropriate

Sites in italics were also proposed for protection in the habitat conservation plan.

network will be established which supports as much of the biodiversity of the Tumbesian Centre of Endemism as possible. Because the extent of endemicity and distributions of the other wildlife groups of the region are so poorly known, it will not be possible to determine how well the proposed reserve network meets their requirements until more research is conducted.

However, in general birds are a good group to base conservation priorities on for the following reasons. They have dispersed to, and diversified in all regions of the world and virtually all habitat types and altitudinal zones (ICBP 1992), and they

Box 8. Priority bird conservation sites in the Tumbesian region
The sites are listed in rank order.

1. North-West Peru Biosphere Reserve (3°49'S 80°17'W), Tumbes Department
Important for 15 priority species.

2. Machalilla National Park (1°36'S 80°42'W), Manabí Province
Important for 14 priority species.

3. Azuay Province site 1 (potentially Manta Real (2°34'S 79°21'W)
For *Pyrrhura orcesi* and *Pachyramphus spodiurus*.

4. Tambo Negro (4°24'S 79°51'W) and/or Celica-Alamor (4°03'S 80°00'W, Loja Province)
For *Syndactyla ruficollis* and *Myrmeciza griseiceps*.

5. Azuay Province site 2 (yet to be identified)
For *Atlapetes pallidiceps* if and when found.

6. El Oro Province site 1 (yet to be identified, potentially Buenaventura (3°40'S 79°44'W)
For *Pyrrhura orcesi*.

play a central role as indicators of environmental change (Carson 1963), reacting quickly to habitat alteration, and have been used for selecting areas for protection in the EEC and for wetlands globally (Fjeldså 1991). They are also easy to observe and survey, so much data can be generated quickly. A literature review conducted by BirdLife International (ICBP 1992, Thirgood and Heath [in press]) established that where avian endemism is pronounced, there appears to be a high degree of endemism in other groups. It should be borne in mind, however, that some groups, especially insects and plants, can show much narrower distribution patterns than birds.

A COMBINED PRIORITY SITE LIST BASED ON HABITAT AND BIRD DATA

In Box 9 we have combined the priority conservation areas in the Tumbesian region based on habitat and bird data into a single list of key sites. This combines both currently protected areas and proposed new reserves and it can be seen that there is a generally good coincidence between the habitat and bird conservation priorities. Until further data are forthcoming from other species groups, this combined list represents the best attempt at an inventory of the most crucial conservation sites in the Tumbesian region.

Box 9. Critical conservation sites in the Tumbesian region based on bird and habitat data. The sites are listed in priority order.

1. North-West Peru Biosphere Reserve
2. Machalilla National Park
3. *Manta Real or alternative Azuay Province humid forest site*
4. *Sozoranga/Catacocha* (equal weighting)
5. *An El Oro Province humid forest site* (potentially Buenaventura)
6. *Hacienda Quesada*
7. *Cabeceras de Bilsa/Jatun Sacha Bilsa*
8. *A second Azuay Province humid forest site*
9. *Celica-Alamor*

Sites in italics are proposed new reserves; the other sites are existing reserves.

CONSERVATION OF OTHER WILDLIFE IN THE TUMBESIAN REGION

The most serious barrier to considering how well the bird and habitat plant conservation recommendations meet the requirements of the other fauna of the region is the lack of biogeographic and distributional data on these other groups. The following is based on the few data that exist.

Mammals

At least four mammal species are restricted to the Tumbesian centre of endemism: the fox *Dusicyon sechurae*, the squirrel *Sciurus stramineus* and two *Phyllotis* mice (Emmons and Feer 1990, Pearson 1982). Virtually nothing is known about the mice, but the fox and squirrel seem to be reasonably tolerant of habitat disturbance (Duckworth 1992). There may be other endemic mammals (especially rodents and bats) yet to be described.

A mammal survey conducted in early 1991 in south-western Ecuador concluded that the degradation, fragmentation and high hunting rates in that region may have been the primary combined causes of impoverished mammal faunas encountered, although more surveys are required to confirm this (Duckworth 1992). It seems likely that those mammal species most dependent on large tracts of unbroken forest (e.g. *Tapirus* spp., some *Felis* spp.) or most susceptible to hunting (e.g. *Cebus* and *Alouatta* monkeys, *Agouti paca*) would benefit most from the protection of the largest and remote forest patches that remain: these are the Machalilla National Park and North-West Peru Biosphere Reserve, which have been given the highest conservation priority above.

The most urgent priority for the mammals of the Tumbesian region is more research to determine if any further endemic species occur and to detail their distributional patterns.

Other fauna

All other groups are so poorly known in the Tumbesian region that intensive surveys are the most measures needed at this stage. The extent to which the proposed reserve network meets the conservation needs of the other wildlife of the region depends on the degree to which their distributions match those of the vegetation types and birds of the Tumbesian region. Particular groups meriting attention are Lepidoptera, reptiles and amphibians. It should be borne in mind that additional areas, not already recognized, may be recommended for protection as data become available from other wildlife groups. However, by using a combination of bird and plant data to design the proposed reserve network, it is hoped that the number of these 'new' areas will be minimal.

AREAS WHICH NEED FURTHER SURVEYS

The selection of areas for conservation action in the above analysis is based on current knowledge of the Tumbesian region. Yet some parts of the Tumbesian region are still very remote and inaccessible, and they could support 'undiscovered' extensive forests, with large numbers of threatened plant and animal species. Satellite images of south-west Ecuador have been carefully checked for such forest patches in preparation for fieldwork in 1989 and 1991 (and also in the preparation of Figure 38). West-central Ecuador and north-western Peru have been less well checked. It is unlikely that any 'undiscovered' forests found in the Tumbesian region will be large. Even if there are, or if future surveys of already-visited sites change their importance (e.g. if an important bird species is found there which had not been seen at the site before), new information can be incorporated into the method used to generate a revised outcome. It should be stressed that conservation action should not wait until these areas have been surveyed.

A few areas which are certainly worthy of future surveys have been identified, they are:

- The Coastal Cordillera of western Ecuador which has several unsurveyed forest patches (at least one of which, the Mache-Chindul area [Mudd 1991], still supports extensive tracts of forest).
- The area of deciduous forest to the west of Macará in southern Loja Province, Ecuador (including the Hacienda Linderos area [Williams and Tobias 1994] and the area between Sabanilla and Zapotillo [recently visited by M. B. Robbins and R. S. Ridgely).
- The Manglares-Churute Ecological Reserve in coastal Guayas Province, which apparently has some fairly extensive forest which has been only briefly surveyed, yet several threatened and near-threatened Tumbesian endemics have already been found there (R. S. Ridgely *in litt.* 1992).
- The Arenillas Military Reserve in El Oro Province, Ecuador, which has extensive deciduous forest which has been only briefly surveyed (Parker and Carr 1992).
- The Cerros de Amotape National Park and the El Angolo Hunting Reserve in Tumbes and Piura Departments, Peru (part of the North-West Peru Biosphere Reserve). These areas may support large tracts of deciduous forest yet they are virtually unknown ornithologically and poorly known botanically.

OTHER ACTION

The conservation measures outlined above have little chance of success if conservation efforts are solely restricted to small areas. In addition to this localized action there must be more widespread environmental education campaigns, along with a regional forest conservation and restoration initiative, and a reforestation programme.

Environmental education

Environmental education campaigns should be carried out by focusing on basic ecological correlations, such as the importance of forests and trees for firewood, as water catchment areas, as nutrient and sediment stores and in preventing soil erosion. Landslides often block roads in the Tumbesian region, often severely disrupting local communications and trade for several days at a time. The role of forests as regulators which contribute to the stability of environments should also be explained and ways of carrying out reforestation should be discussed.

Municipal leaders in both Sozoranga and Celica in Loja Province, Ecuador have requested such materials for schools and community projects. The Loja-based NGO ArcoIris, which has produced similar materials for the Podocarpus National Park, are planning to produce materials in these areas (M. Kelsey *in litt.* 1993). This would involve close collaboration with school teachers and other community figures. Reserves close to such communities represent excellent resources and the importance of forest to protect water supplies is apparently widely recognized in this semi-arid region.

Local involvement

People living close to the new reserves must be fully informed of the purpose of the reserve and the benefit it can bring them; they should be encouraged to participate in their management. Colourful posters promoting forests could easily be provided and the local newspapers and radio should be used as much as possible. Only a large-scale publicity campaign will get the message through to all the local people and slowly they will begin to accept that conservation is not something imposed on them by outsiders, but rather it is something important for themselves. Hopefully a feeling of pride in the importance of each community's local forest would develop, as had occurred at Sozoranga in Loja Province, where repeated radio and newspaper coverage of the international importance of the forests in that region have raised greatly local awareness of ecology and conservation. A national environmental education programme has been submitted to the Ecuadorian government (W. Oliver verbally 1991); such schemes should be supported by international conservation organizations.

Efforts must be made to ensure that employment opportunities in newly created protected areas are directed towards the local people so that they feel that they are benefiting from the conservation action. There are opportunities for ecotourism in the protected area network proposed, and local people should be trained and employed as guides. This benefit, however, must not be overstressed to avoid rasing false expectations. Revenue from ecotourism should, where possible, also be put back into areas beneficial to local people.

Reforestation

The regional forest conservation and reforestation programme poses the biggest logistical problems. One of the worst problems with current land-use in the Tumbesian region is the grazing of cattle inside forest. This eventually destroys the forests (mainly through the prevention of natural regeneration) and degrades the understorey to the detriment of the flora and fauna. It also produces poor grazing areas. Therefore, the strict separation of grazing areas, crop-growing land and forest areas would be important for the development of long-term sustainable land-use. Only then can optimal land-use techniques be developed for each of the different land-use types. Ideally grazing should be excluded from all remaining forests as this may be the most important factor leading to their degradation and lack of recruitment of the remaining forests (forest clearance for agriculture is the next most important factor; wood cutting is of limited importance).

As in many developing regions, people in the Tumbesian region are often economically and nutritionally dependent on their cattle, especially in areas which are no longer suitable for crop cultivation, or where the growing season is very short. In such instances it is not possible to expect people to give up cattle grazing inside forests. Forests in the Tumbesian region (unlike in many Andean areas) often seem to have little value to people. Many local homes have electricity, and fuelwood is only locally in short supply. In some areas of northern Peru wood (especially *Prosopis*) is used for the fabrication of charcoal; in this case the forests have some economic value. However, in most of the region forests are only used for animal-grazing, hunting, gathering a few medicinal plants, fruits and nuts. Soils are often good, so profitable agriculture can be carried out in many areas which are not too steep or too dry. Generally it is difficult to measure the indirect value of forests, such as their role in effective watershed protection or in preventing soil erosion, versus the direct economic value people gain from cutting them down and planting crops. Convincing local people of the value of forests will be one of the most significant challenges.

Alternatives to forest destruction

Studies of alternative techniques to land management should be made; this could be carried out by pilot agricultural projects on agroforestry and model farms to demonstrate the benefit of sustainable techniques. A reforestation project in the Machalilla National Park could be used as a model for other similar projects. Local tree species should be planted in preference to ecologically damaging eucalyptus and pines.

The need for a symposium on the Tumbesian region

The regional environmental plan may best be produced through a regional workshop of Governmental Agencies, NGOs and community leaders. The most appropriate people to produce such a plan are those who can carry the responsibility for its implementation and those most affected by it.

The workshop should aim to produce a practical guide on how to tackle the environmental problems of the region and a plan for actions to be taken to protect the remaining forest of the Tumbesian region. Working groups should discuss particular topics (such as how to get the acceptance and involvement of local people; what reforestation techniques are suitable; the legal background to any programmes).

It is hoped that the recommendations included here will form the basis of a plan prepared by Ecuadorian and Peruvian national conservation organizations and NGOs in partnership with the local people of the Tumbesian region, and endorsed by the governments of the two countries. If, as we suggest, the conservation plan combines ecologically sound measures with socially desirable actions the likelihood of success will be improved.

PRIORITIES FOR RESEARCH

Although habitat protection is the highest conservation priority, regional research must continue so that the biological database on the Tumbesian region can be enlarged. Sixteen more urgent topics have been identified below. These refer principally to work in the fields of botany and ornithology. It should be noted that given the current state of knowledge it is not possible to give as detailed recommendations for botanical research as for ornithological work and any further botanical survey would provide valuable information.

1. Mapping the vegetation cover of the Tumbesian region accurately

The extent of forest remaining in the region is not precisely known, as data gathered to date only cover some areas and habitat types. While humid forests are usually discernible on LANDSAT satellite images, this is not the case for dry forests, which can easily be confused with scrub or even agricultural areas. Also, the state of degradation of dry forests cannot be assessed on satellite images, and extensive cloud cover can often obscure the images. Therefore future botanical surveys should concentrate on dry forests yet unsurveyed. Most important among these are the hills west of Zapotillo in Loja Province, Ecuador, and the dry forests on the foothills in Piura and Lambayeque Departments, Peru. Other areas which need more surveys are the arid intermontane valleys of Azuay and Loja Provinces in Ecuador, and Piura Department in Peru, in order to find patches of Deciduous to Semi-evergreen Lower Montane Cloud-forest and Intermontane Thorn-forest and Forest. In all these dry forests ground visits are necessary to determine the state of the understorey and the amount of tree regeneration.

Humid forests are more easily mapped from satellite images and from aeroplanes and therefore are already better known than the dry forests of the Tumbesian region. However, some areas deserve special attention and should, if possible, be surveyed with overflights. These areas are the very wet Andean foothills of El Oro and Azuay Provinces, Ecuador, and the Coastal Cordillera to the north-west of the Machalilla National Park.

While the value of such an exercise cannot be stressed too greatly, it has to be feared that no large, previously unknown forest patches will be found. Therefore it would be wrong to delay urgent conservation action until a complete coverage of the remaining forest cover is available. Instead, the value of such a database will lie in the possibility of accurately monitoring future forest destruction. The total population size of each priority bird species (and other fauna) could be estimated by combining density estimates with knowledge of the extent of the habitat that particular species occupy.

2. Unsurveyed or under-surveyed areas should be visited

Figure 45 (page 133) showed that the ornithological survey effort in the Tumbesian region to date has been uneven: most work has been carried out in south-west Ecuador (especially El Oro and Loja Provinces). Particular areas meriting further study are highlighted on page 194. At certain sites which have had limited surveys, some periods of the year are completely unknown (see below). Unsurveyed but potentially suitable sites should be identified by aerial photography as described above.

3. More survey work to build up a year-round database on the distributions of priority Tumbesian birds

Figure 88 illustrates the known ornithological effort by month since 1970 at selected sites in the Tumbesian region. It shows the months during which surveys are most urgently needed. Survey effort has been far from even, with two peaks:

Figure 88. Ornithological survey effort by month at selected sites in the Tumbesian region.

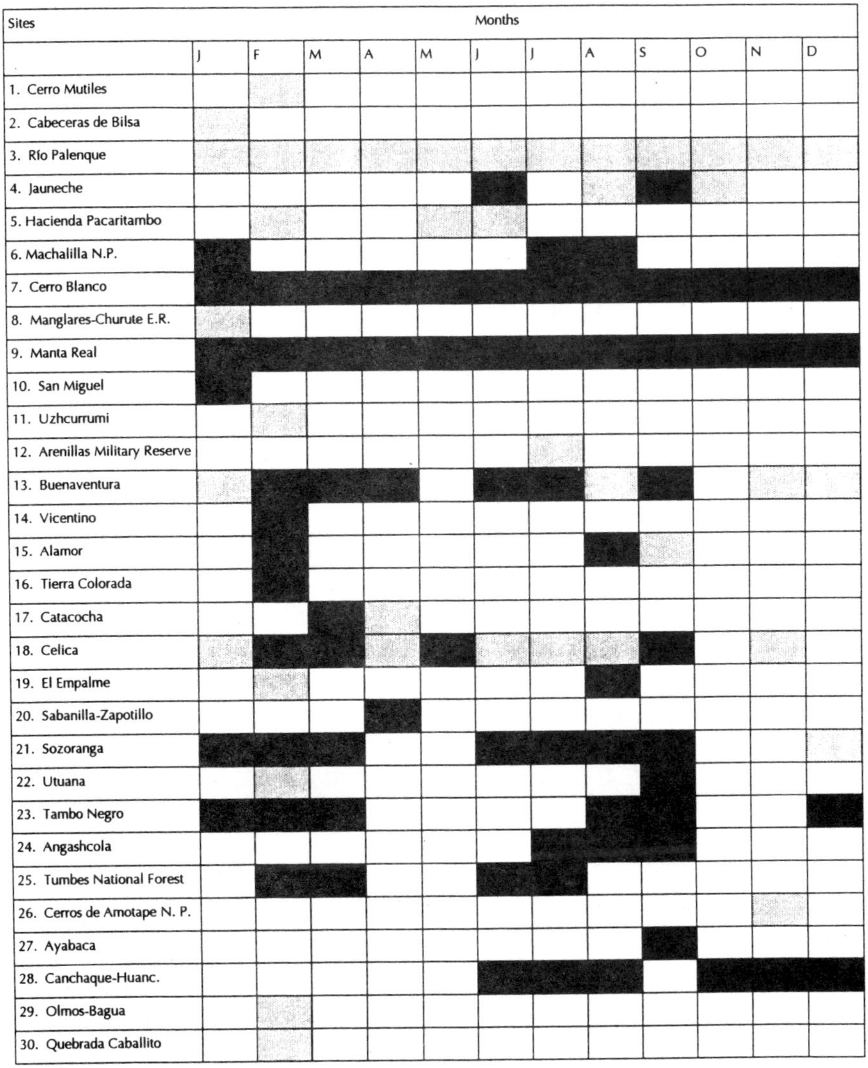

Dark shading indicates surveys of five days or more duration; light shading suveys less than five days. Only surveys for which dates were available are shown.

Sources: Vuilleumier (1978), Leck (1979), Leck *et al.* (1980), Schulenberg and Parker (1981), Wiedenfeld *et al.* (1985), P. J. Greenfield *in litt.* to ICBP (1989), R. S. Ridgely *in litt.* to ICBP (1989), Best and Clarke (1991), Bloch *et al.* (1991), P. Coopmans *in litt.* (1991), M. Kessler *in litt.* (1992), Kiff (1991), Krabbe (1991), Ridgely (1991a, 1991b), R. S. Ridgely *in litt.* (1991), Best (1992), P. Coopmans *in litt.* (1992), N. Krabbe *in litt.* (1992), M. B. Robbins *in litt.* (1992), Parker and Carr (1992), M. Whittingham *in litt.* (1992), K. S. Berg *in litt.* (1993), M. Kessler *in litt.* (1993), Williams and Tobias (1994) and Parker *et al.* (1995).

January to March and July to September. There are some months (e.g. May, October to December) when virtually no surveys have been conducted, and at several sites (e.g. Jauneche, Arenillas Military Reserve) survey work has been confined to the dry (non-breeding) season. Future workers in the Tumbesian region are urged to time their visits to maximize survey time during 'new' months.

4. Studies of the population size of each threatened and near-threatened Tumbesian bird species

Quantitative studies of the population size of each priority species should be conducted over several months in the dry and wet seasons at as many sites as possible. The total population size of each priority species can be established as explained above. Knowledge of population sizes is important if the degree of threat to each is to be established.

5. Detailed ecological studies of the threatened and near-threatened Tumbesian bird species

Very few details are known about the habitat requirements, feeding preferences and life-histories of these species. Such data can only be gathered by quantitative ecological studies lasting several months at least, in both the wet and dry seasons. A topic of central importance is the degree to which priority species can tolerate degradation, and can withstand and survive in the long term under various human activities taking place in the region. Six species (*Leptotila ochraceiventris, Synallaxis tithys, Syndactyla ruficollis, Hylocryptus erythrocephalus, Myrmeciza griseiceps* and *Lathrotriccus griseipectus*) appear especially sensitive to understorey degradation, chiefly by grazing animals. This should be investigated by quantitative studies of sites with varying states of understorey degradation during the dry and wet season to establish the presence or absence of these understorey species.

Several non-understorey dependent species (e.g. *Leucopternis occidentalis* and *Ortalis erythroptera*) occur in largely cleared, agricultural areas close to forest remnants; the nature of their dependence on these forest fragments for breeding is, however, unclear at present and requires further investigation. Another topic which requires study is the seasonality of the priority species' habitat needs, including their need for different habitats and different altitudes. This could be achieved by radio-tracking or ringing species of particular interest. Data gained from these studies can be used to design management programmes for protected areas, and in environmental education schemes.

6. Studies of the seasonal movements of Tumbesian species

There may be pronounced seasonal movements of Tumbesian species between habitats. Seasonal movements of priority Tumbesian species can be further investigated as the year-round database of bird records from sites in the region becomes established. A more direct approach would be to radio-track or ring target species.

7. Studies of the interannual variations in the distribution of Tumbesian species

There is also potential for interannual variations in the distributions and population sizes of Tumbesian species. The climate (especially rainfall) of the region is typified by pronounced fluctuations from year to year (see 'Climate' section and Munday and Munday 1992) which has consequences for the vegetation, especially the deciduous types. Very wet El Niño years complicate the issue. The only study from which there are data from several successive years (Marchant 1958) found marked interannual fluctuations in both the distribution of birds and the time of the breeding season on the Santa Elena Peninsula in western Ecuador. Further long term studies are required which employ standardized methodologies.

8. Studies of the extent and sustainability of the international parrot trade

Both *Aratinga erythrogenys* and *Brotogeris pyrrhopterus* are in international trade and the large numbers reported to CITES in recent years have caused alarm among conservationists. Between 1983 and 1988 there were 51,853 *A. erythrogenys* and 59,320 *B. pyrrhopterus* reported to CITES as in trade (Inskipp and Corrigan 1992). Peru does not permit trade in its Amazonian species and it is possible that to allow these to enter trade they are declared as *A. erythrogenys* or *B. pyrrhopterus* and exported legally. The extent of this mis-declaring of parrots exported from Peru and the prevalence of illegal smuggling of these species from Ecuador (where they are officially protected) to Peru should be investigated. The first by liaison with the customs officials in Peru to establish the extent of any mis-declaring, and the second by more efficient monitoring of cross-border smuggling by the customs officials at the military checkpoints on the Peru-Ecuador border, in addition to discussions with parrot traders.

Data provided by studies of the population size and breeding success of the parrots, combined with CITES/TRAFFIC data, will help to establish whether the trade figures are accurate, and the trade itself is sustainable. It will also facilitate an assessment of their status, at present very difficult to make for a variety of reasons (Best *et al.* in press).

9. Studies of the impact of hunting

At present, the extent and impact of hunting on the priority species is unknown.

Crypturellus transfasciatus, *Penelope albipennis* and *Ortalis erythroptera* are all hunted for food and other species (e.g. *Leucopternis occidentalis*) for sport. Hunting should be investigated by interviewing local people. Regional or seasonal trends should be assessed, along with the reasons for the birds being hunted.

10. Floristic surveys

Far too little is known about the floristic composition of the Tumbesian vegetation types. Only two localities (Capeira and Jauneche) have so far been completely covered; several other areas (e.g. Puná Island, the Santa Elena Peninsula, the Colonche-Chongón Cordillera and parts of the Machalilla and Cerros de Amotape National Parks) have been studied in some detail, while most of the region has only been subject to opportunistic collecting, making the Tumbesian region one of the least known botanically in Peru and Ecuador. Vegetation types which are particularly little-known are the Semi-evergreen Lowland to Lower Montane Forests, the Humid to Very Humid Premontane Cloud-forest, the Deciduous to Semi-evergreen Intermontane Scrub, Thorn-forest and Forest, the Humid Lower Montane Cloud-forest and the Deciduous to Semi-evergreen Lower Montane Cloud-forest. Selected localities of these forest types should be sampled in detail, covering trees, understorey plants, herbs and epiphytes.

11. Production of a more detailed vegetation classification

The vegetation classification presented earlier in this book is simply an initial starting point upon which a complete coverage of the vegetation types of the Tumbesian region can be built after more detailed surveys. These will also provide a basis for a more accurate qualitative comparison of the vegetation types, and for a subdivision of those vegetation types which currently do not fit in well with the proposed classification (especially in the case of the Semi-evergreen to Evergreen Lowland to Montane Forest).

12. Long-term studies on the ecology and phenology of the different vegetation types, especially in connection with migration patterns of birds and insects

Long-term ecological studies of the Tumbesian forests are at present completely lacking, so nothing is known about the phenology of the plants as species and as individuals, the relationships of the vegetation to climatic parameters, or the relations of the plants to insect, bird and bat pollinators and dispersers. So far, only the studies conducted over the last few decades outside the region in the moist forest on Barro Colorado Island, Panama, and the dry forest in the Santa Rosa National Park, Costa Rica, provide clues to the nature of these relationships.

This situation provides an excellent opportunity to compare the Tumbesian forests with similar Central American forest types. The possibilities for meaningful and interesting ecological research are immense. From a conservation point of view, questions relating to the seasonal and interannual migration of animal species should have the highest priority, since the survival of many animal species might depend on forest patches which are only used for a short part of the year or in exceptional years (e.g. El Niño events).

Certainly the most suitable areas to establish field stations for long-term studies are the North-West Peru Biosphere Reserve and the Machalilla National Park, since they contain the largest and best preserved forest areas in the Tumbesian region and provide access to several adjacent but ecologically very different vegetation types.

13. Studies of the population dynamics of commercial tree species

These should provide a database on which sustainable forest use might eventually be developed. Respective research on the sustainability of the use of other forest products (both plant and wildlife) is also needed.

14. Studies of the use of native plant species and of reforestation techniques

A more practical line of research would include ethnobotanical studies and especially the development and implementation of reforestation techniques with native tree species. This will be of increasing importance as fuelwood resources diminish and erosion threatens the destruction of valuable soil. In order to increase the acceptance of the project a method of reforestation which benefits the local inhabitants from the start is preferred over one that takes 10 or 20 years. Comparable efforts in Costa Rica should provide a useful starting point for this research.

15. Intensive surveys of all wildlife groups occurring in the Tumbesian region

In addition to the proposed research on birds and flora outlined above, each other wildlife groups should be further studied. For some groups (e.g. mammals) small amounts of data exist, whereas for others (e.g. reptiles, insects, fungi) very few surveys have been conducted. Studies should concentrate on endemicity and distributional patterns. Such studies are vital to determine whether the proposed conservation measures also protect these less well known groups. Research should be carried out first in the areas known to be important for plants and birds.

16. Socio-economic studies of the people of the Tumbesian region

Key elements of research are socio-economic studies of the people adjacent to (or in some cases within) the existing and proposed protected areas in order to determine the ways in which they use these areas. Studies would help determine the way in which wildlife is perceived and the ways in which people benefit from, and come into conflict with wildlife.

Box 10. Summary of conservation action needed in the Tumbesian region.

SUPPORT FOR EXISTING PROTECTED AREAS

Machalilla National Park and the North-West Peru Biosphere Reserve must be securely protected and effectively managed by:
- creation of effective buffer zones
- increased staff and resources
- creation of research stations
- employment of local people as guides/wardens
- environmental education in and around the parks

The private reserves of the region, notably Cerro Mutiles, Río Palenque, Jauneche and Cerro Blanco should be supported so that each plays it full role in regional conservation.

FURTHER RESEARCH

Additional surveys should be undertaken in several areas, notably the Manglares - Churute Ecological Reserve and the southern part of the North-West Peru Biosphere Reserve.

Many individual research priorities remain, involving bird and habitat surveys, ecological research and socio-economic studies.

RESERVE CREATION

Seven new reserves should be created in Ecuador to protect the habitats and birds of the Tumbesian region. They should be at:
- Manta Real or another Azuay Province humid forest site
- Sozoranga or Catacocha in Loja Province
- a humid forest site in El Oro Province
- Hacienda Quesada in Azuay Province
- Cabeceras de Bilsa in Esmeraldas Province
- a second humid forest site in Azuay Province
- Celica-Alamor in Loja Province

Funds should be made available for the purchase of each area, but also the longer term management and protection of the sites.

REGIONAL ENVIRONMENTAL EDUCATION AND AWARENESS PROGRAMME

This should include:

- education in the value and importance of forests
- local involvement in the management of new reserves
- training in the reforestation of degraded areas for sustainable use, providing alternatives to forest destruction

A workshop bringing together diverse groups should produce a comprehensive environmental plan for the region.

SUMMARY OF CONSERVATION ACTION NEEDED

The conservation recommendations we have presented for the Tumbesian region can be split into four main elements: the reinforcement of currently protected areas; the creation of new reserves; a regional environmental education and awareness programme; and an ongoing research effort (Box 10). It is important that each element is implemented simultaneously.

Finally we stress that these conservation measures represent merely a starting point upon which a much more comprehensive and detailed programme can be developed. This can only be done when all those with an interest in biodiversity conservation in the region are brought together, within Ecuador or Peru, in order to prepare a conservation programme for the Tumbesian region.

■

REFERENCES

ACKERY, P.R. AND VANE-WRIGHT, R. I. (1984) *Milkweed butterflies*. London: British Museum (Natural History).

ACOSTA-SOLIS, M.A. (1959) *Los Manglares del Ecuador.* Quito: Ecuadorian Institute of Natural Sciences.

ACOSTA-SOLIS, M.A. (1968) *Divisiones fitogeográphicas y formaciones geobotánicas del Ecuador.* Quito: Casa de la Cultura Ecuatoriana.

ACOSTA-SOLIS, M.A. (1977) *Ecología y Fitoecología.* Quito: Casa de la Cultura Ecuatoriana.

ANDRADE, G. I. AND MEJIA, C. (1991) Cambios estacionales en la distribución de la avifauna terrestre en el Parque Nacional Natural Macuira, Guajira, Colombia. *Trianea* 1: 145-169.

AMERICAN ORNITHOLOGISTS' UNION (AOU) (1983) Check-list of North American birds. Kansas, U.S.A.: Allen Press.

AOU (1987) 36th supplement to the AOU check-list of North American birds. *Auk* 104: 591- 596.

AOU (1989) 37th supplement to the AOU check-list of North American birds. *Auk* 106: 532-538.

ARRIAGA, L. (1987) Manejo de recursos costeros en el Ecuador. Pp. 3-10 In *La Pesca Artesanal en el Ecuador.* Quito, Ecuador: EPSOL, CEPLAES, ILDIS.

ATLAS DEL ECUADOR (1982). Paris. j.a.

BALSLEV, H. (1988) Distribution patterns of Ecuadorean plant species. *Taxon* 37: 567-577.

BEST, B. J. AND CLARKE, C. T. EDS (1991) *The threatened birds of the Sozoranga region, south-west Ecuador.* Cambridge, U.K.: International Council for Bird Preservation (Study Report 44).

BEST. B. J. ED. (1992) *The threatened forests of south-west Ecuador.* Leeds, U.K.: Biosphere Publications.

BEST, B. J., BROOM, A. L., CHECKER, M. AND THEWLIS, R. M. (1992) An ornithological survey of El Oro and western Loja Provinces, south-west Ecuador, January-March 1991. Pp. 137-210 in B. J. Best, ed. *The threatened forests of south-west Ecuador.* Leeds, U.K.: Biosphere Publications.

BEST, B. J., CLARKE, C. T., CHECKER, M., BROOM, A. L., THEWLIS, R. M., MCNAB, A. AND DUCKWORTH, J. W. (1993) Distributional records, natural history notes and conservation of some poorly-known birds from south-western Ecuador and north-western Peru. *Bull. Brit. Orn. Club.* 113: 108-120; 234-255.

Best, B. J. and Krabbe, N. (1994) A review of the status and conservation of the Rufous-headed Chachalaca. *World Pheasant Assoc. J.* 17-18:45-56.

Best, B. J., Krabbe, N., Clarke, C. T. and Best, A. L. (in press) Red-masked Parakeet *Aratinga erythrogenys* and Grey-cheeked Parakeet *Brotogeris pyrrhopterus*: two threatened parrots from Tumbesian Ecuador and Peru? *Bird Conserv. Internatn.*

Blake, J. G. and Loiselle, B. A. (1991) Altitudinal migrants at Estación Biologica La Selva and Parque Nacional Braulio Carrillo, Costa Rica. *Paper given at 2nd congress on Neotropical ornithology, Quito.*

Bloch, H., Poulsen, M. K., Rahbek, C. and Rasmussen, J. F. (1991) *A survey of the montane forest avifauna of the Loja Province, southern Ecuador.* Cambridge, U.K.: International Council for Bird Preservation (Study Report 49).

Bond, J. (1980) *Birds of the West Indies.* Boston: Houghton Mifflin.

Borhidi, A. (1976) *Fundamentals of the geobotany of Cuba.* Budapest: The Hungarian summary of D.Sc. thesis.

Borhidi, A. (1991) *Phytogeography and vegetation ecology of Cuba.* Budapest: Akademiai Kiodo.

Brandbyge, J. (1992) Planting of local woody species in the páramo. Pp. 265-274 in H. Balslev and J. L. Luteyn, eds. *Páramo: an Andean ecosystem under human influence.* London: Academic Press.

Braun-Blanquet, J. (1921) Prinzipien einer systematik der pflanzengesellschaften auf floristischer grundlage. *Naturw. Ges.* 57: 305-350.

Brittanica World Data (1991) The Encyclopaedia Brittanica supplement.

Brosset, A. (1964) Les oiseaux de Pacaritambo (Ouest de l'Ecuador). *Oiseaux et R.F.O.* 324: 1-24; 112-135.

Brown, K. S. (1982) Palaeoecology and regional patterns of evolution of Neotropical forest butterflies. Pp. 255-308 In G. T. Prance, ed. *Biological diversification in the tropics.* New York: Columbia University Press.

Brown, K. S. (1987) Biogeography and evolution of Neotropical butterflies. Pp. 66-104 in T. C. Whitmore and G. T. Prance, eds. *Biogeography and Quaternary history in tropical America.* Oxford: Clarendon Press.

Bush, M. B., Colinvaux, P., Wiemann, M. C., Piperno, D. R. and Liu, K. B. (1990) Late Pleistocene temperature depression and vegetation change in Ecuadorian Amazonia. *Quat. Res.* 34: 330-345.

Campbell, K. E. (1982) Late Pleistocene events along the coastal plain of northwestern South America. Pp. 423-440 In G. T. Prance, ed. *Biological diversification in the tropics.* New York: Columbia Univ. Press.

Cañadas Cruz, L. (1983) *El mapa bioclimático y ecológico.* Quito: Banco Central del Ecuador.

Cañadas Cruz. L. and Estrada, W. (1978) *Ecuador - mapa ecológico.* Quito:

PRONAREG-Ecuador, Ministerio de Agricultura y Ganaderia.

Carsen, R. (1963) *Silent spring*. London: Penguin.

CDC-UNALM (1992) *Estado de conservación de la diversidad natural de la región noroeste del Perú*. Lima: Centro de Datos de Conservación.

Chapman, F. M. (1926) The distribution of birdlife in Ecuador. *Bull. Am. Mus. Nat. Hist.* 56: 1-784.

Cintrón, G. (1984) Criaderos comerciales del camarón en el Ecuador. In *Recopilación sobre la conservación, aprovechamiento y manejo del manglar*. Quito: MAG Informe 33.

Clarke, C. T. (1990) New populations of rare cracids found in Ecuador and Peru. *World Pheasant Association News* 28: 6-9.

Clüsener, M. and Breckle, S. W. (1987) Reasons for the limitation of mangrove along the west coast of northern Peru. *Vegetatio* 68: 173-177.

Colinvaux, P. (1987) Amazon diversity in the light of the paleoecological record. *Quat. Sci. Rev.* 6: 93-114.

Collar, N. J. and Andrew, P. (1988) *Birds to watch: the ICBP checklist of threatened birds*. Cambridge, U.K.: International Council for Bird Preservation (Techn. Publ. 8).

Collar, N. J., Gonzaga, L. P., Krabbe, N., Madroño Nieto, A., Naranjo, L. G., Parker, T. A. and Wege, D. (1992) *Threatened birds of the Americas*. Cambridge, U.K.: International Council for Bird Preservation. (ICBP and IUCN Red Data Book).

Collar, N. J., Crosby, M. J. and Stattersfield, A. J. (1994) *Birds to Watch 2: the world list of threatened birds*. Cambridge, U.K.: BirdLife International (Conservation Series no. 4).

Connell, J. H. (1978) Diversity relations in Cape shrublands and other vegetations in the South-eastern Cape, South Africa. *Vegetatio* 54: 103-127.

Cracraft, J. (1985) Historical biogeography and patterns of differentiation within the South American avifauna: areas of endemism. Pp. 49-84 in P. A. Buckley, M. S. Foster, E. S. Morton, R. S. Ridgely and F. G. Buckley, eds. *Neotropical ornithology*. Washington, D.C.: American Ornithologists' Union. (Orn. Monogr. 36).

Cracraft, J. and Prum, R. O. (1988) Patterns and processes of diversification: speciation and historical congruence in some Neotropical birds. *Evolution* 42: 603-620.

Croat, T. and Busey, P. (1975) Geographical affinities of the Barro Colorado Island flora. *Brittonia* 27: 127-135.

Diels, L. (1937) Beiträge zur kenntnis der vegetation und flora von Ecuador. *Bibl Bot.* 29: 1-190.

Dillon, M., Sagásleguí, A., Sánchez, I. and Llatas, S. (1993). Floristic inventory and biographic analysis of montane forests in northern Peru. Pp. 33 in H. Balslev, ed. *Neotropical montane forests: biodiversity and conservation.*

Univ. of Aarhus, Denmark: AAU Reports 31.

Dodson, C. and Gentry, A. H. (1978) Flora of the Río Palenque Science Centre, Los Ríos, Ecuador. *Selbyana* 4: 1-628.

Dodson, C., Gentry, A. H. and Valverde, F. M. (1985) Flora of Jauneche. *Selbyana* 8: 1-512.

Dodson, C. and Gentry, A. H. (1991) Biological extinction in western Ecuador. *Ann. Miss. Bot. Gard.* 78: 273-295.

Dodson, C. and Gentry, A. H. (in press.) *Flora of Capeira.*

Dourojeanni, M. (1986) Recursos naturales, desarrollo y conservación en el Peru. In M. Baca, ed. *Gran Geográfia del Peru.* Vol. 4. Madrid.

Duckworth, J. W. (1992) Mammals found in south-west Ecuador, January-March 1991. Pp. 121-134 in B. J. Best, ed. *The threatened forests of south-west Ecuador.* Leeds, U.K.: Biosphere Publications.

Ehlers, K. and Ehlers, R. (1990) *Tillandsias of Ecuador, Provinces El Oro, Azuay, and Loja.* Stuttgart. Unpublished manuscript.

Ellenberg, H. (1975) Vegetationstufen in perhumiden bis perariden bereichen der tropischen Anden. *Phytocoenologia* 2: 368-387.

Emmons, L. and Feer, F. (1990) *Neotropical rainforest mammals: a field guide.* Chicago and London: University of Chicago Press.

Empraire, L. and Friedberg, C. (1990) *Releves floristiques des regions de Piura (Perou) et de Loja (Equateur).* Paris: Convention ORSTOM - PUC.

FAO (1982) *Food and agriculture legislation* 31: 74-99.

FAO-UNESCO (1988) *FAO-UNESCO Soil map of the world, revised legend.* Rome: FAO (World Soil Res. Rep. 60).

Figueroa, S. (1992) Patrimonio de áreas naturales y otras áreas naturales en Ecuador. Pp. 207-222 In S. Amend and T. Amend, eds. *¿Espacios sin habitantes? Parques Nacionales de América del Sur.* Caracas, Venezuela: IUCN and Editorial Nueva Sociedad.

Fjeldså J. (1991) The conservation of biological diversity: using birds to set priorities. Pp. 157-175 in J. Hansen, ed. *Environmental concerns, an interdisciplinary exercise.* London: Elsevier.

Fjeldså, J. and Krabbe, K. (1990) *Birds of the high Andes.* Copenhagen: Apollo Books.

Fjeldså, J. (1992) Biogeographic patterns and evolution of the avifauna of relict high-altitude woodlands of the Andes. *Steenstrupia* 18: 9-62.

Fjeldså, J. and Kessler, M. (in press) A strategy for preserving the biological diversity of the *Polylepis* woodlands of the high Andes in Peru and Bolivia. Copenhagen: Zoological Museum.

Flenley, J. R. (1979) *A geological history of tropical rainforests.* London: Butterworths.

Fundacíon Natura and SUFOREN (1992) *Parques Nacionales y otras áreas naturales protegidas del Ecuador.* Quito: Fundación Natura.

GENTRY, A.H. (1977) Endangered plant species and habitats of Ecuador and Amazonian Peru. In G. T. Prance, and T. S. Elias, eds. *Extinction is forever.* New York Botanic Garden.

GENTRY, A.H. (1978) Floristic knowledge and needs in Pacific tropical America. *Brittonia* 30: 134-153.

GENTRY, A. H. (1982) Neotropical floristic diversity: phytogeographical connections between Central and South America, Pleistocene climatic fluctuations, or an accident of the Andean orogeny? *Ann. Missouri Bot. Gard.* 69: 557-593.

GENTRY, A. H. (1988) Changes in plant diversity and floristic composition on environmental and geographical gradients. *Ann. Missouri Bot. Gard.* 75: 1-34.

GENTRY, A. H. (1989) Speciation in tropical forests. Pp. 113-134 in L. B. Holm-Nielsen, I. C. Nielsen and H. Balslev, eds. *Tropical forests.* London: Academic Press.

GENTRY, A.H. (1992) Diversity and composition of Neotropical dry forests. Unpublished manuscript.

GOODWIN, D. (1970) *Pigeons and doves of the world.* London: Trustees of the British Museum (Natural History).

GILMARTIN, A. (1972) *The bromeliaceae of Ecuador:* Phanerogamarum Monographiae 4, J. Germany: Cramer, Vaduz

GRAF, K. (1986) Klimaund vegetationsgeographie der Anden. *Physische-geographie.* Zürich: Geogr. Inst. Univ. Zürich. Vol. 19.

HAFFER, J. (1969) Speciation in Amazonian forest birds. *Science* 165: 131-137.

HAFFER, J. (1974) Avian speciation in tropical South America. *Publ. Nuttall Ornithol. Club* No. 14.

HAFFER, J. (1982) General aspects of the refuge theory. Pp. 6-24, in G. T. Prance, ed. *Biological diversification in the tropics.* New York: Columbia University Press.

HAFFER, J. (1985) Avian zoogeography of the Neotropical lowlands. Pp. 113-146 in P. A. Buckley, M. S. Foster, E. S. Morton, R. S. Ridgely, and F. G. Buckley, eds. *Neotropical ornithology.* Washington D.C.: American Ornithologists' Union. (Orn. Monogr. 36).

VAN DER HAMMEN, T. (1981) The Pleistocene changes of vegetation and climate in the northern Andes. Appendix IV. Pp. 125-145 in S. Hastenrath, ed. *The glaciations of the Ecuadorian Andes.* Rotterdam: A. A. Balkema.

HARLING, G. (1979) The vegetation types of Ecuador - a brief survey. Pp. 165-174 in K. Larsen, and L. B. Nielsen, eds. *Tropical botany.* London: Academic Press.

HARLING, G. (1986) Current Scandinavian botanical research in Ecuador *Rep. Botanical. Institute, Univ. Aarhus* 15: 9-10.

HARLING, G. AND SPARRE, B. EDS. (since 1979) *Flora of Ecuador.* Göteborg, Sweden.

HASSAUREK, F. (1967) *Four years among the Ecuadorians (1861-1866).* Southern

Illinois University. Carbondale and Edwardsville. (cited in Dodson and Gentry 1991).

HILTY, S. L. AND BROWN, W. L. (1986) *A guide to the birds of Colombia.* Princeton, U.S.A.: Princeton University Press.

HOLDRIDGE, L. R. AND LITTLE, E. L. (1944) *The forests of western and central Ecuador.*

HOLDRIDGE, L. R. (1956) Determination of world plant formations from simple climatic data. *Science* 105: 367-368.

HOLDRIDGE, L. R. (1967) *Life zone ecology.* San José, Costa Rica: Tropical Science Centre.

HOLDRIDGE, L. R. (1971) *Forest environments in tropical life zones, a pilot study.* Oxford, U.K .: Pergamon.

HOOGHIEMSTRA, H. (1984) *Vegetational and climatic history of the high plain of Bogota, Colombia: A continuous record of the last 3.5 million years.* Cramer, Vaduz.

HUBBELL, S. P. (1984) Methodologies for the study of the origin and maintenance of tree diversity in tropical rain forest. Pp. 8-13 In G. Laury-Lechon, M. Hadley and T. Younes, eds. *The significance of species diversity in tropical forest ecosystems.* Biology International (IUBS). Special issue 6.

HUBBELL, S. P. AND FOSTER, R. (1986) Biology, chance and history in the structure of tropical rain forest tree communities. Pp. 314-329 In J. Diamond and T. Case, eds. *Community ecology.* New York: Harper and Row.

HUECK, K. AND SEIBERT, P. (1981) *Vegetationskarte von Südamerika.* Stuttgart: Gustav Fischer Verlag.

HUMBOLDT, A. V. AND BONPLAND, A. (1807) *Ideen zu einer Geographie der Pflanzen.* Tübingen, Germany (1807), reprinted Darmstadt 1963.

ICBP (1992) *Putting biodiversity on the map: priority areas for global conservation.* Cambridge, U.K.: International Council for Bird Preservation.

INSKIPP, T. P. AND CORRIGAN, H. *Review of significant trade in animal species included in CITES Appendix II.* Cambridge, U.K.: World Conservation Monitoring Centre/International Union for the Conservation of Nature and Natural Resources.

IUCN (1982) *IUCN directory of Neotropical protected areas.* Gland, Switzerland: International Union for the Conservation of Nature and Natural Resources.

IUCN (1988) *Ecuador: conservation of biological diversity and forest ecosystems.* Cambridge, U.K.: Conservation Monitoring Centre.

IUCN (1990) *List of the threatened plant species of Ecuador.* Unpublished.

IUCN (1992) *Protected areas of world: a review of national systems, 4: Nearctic and Neotropical.* Gland, Switzerland and Cambridge, U.K.: International Union for Conservation of Nature and Natural Resources.

JANZEN, D. H. (1986) The future of tropical ecology. *Ann. Rev. Ecol. Syst.* 17: 305-

324.

JANZEN, D. H. (1988) Tropical dry forests: the most endangered tropical ecosystem. Pp. 130-137 in E. O. Wilson, ed. *Biodiversity*. Washington, DC.: National Academy Press.

JANZEN, D.H. (1988) Management of habitat fragments in a tropical dry forest: growth. *Ann. Missouri Bot. Gard.* 75: 105-116.

JORDAN, E. (1988) Die mangrovenwälder Ecuadors im spannungsfeld zwischen ökologie und ökonomie. Pp. 97-138 in H J. Bucholz and G. Gerold, eds. *Lateinamerikaforschung*. Jahrb. der Geogr. Ges. Hannover.

JØRGENSEN, P. M., ULLOA U.C., VALENCIA, R. AND MADSEN, J. E. (1993) Floristic analysis of high-montane Ecuador. Pp. 56 In H. Balslev, ed. *Neotropical montane forests: biodiversity and conservation.* Univ. of Aarhus, Denmark: AAU Reports 31.

KESSLER, M. (1992) The vegetation of south-west Ecuador. Pp. 79-100 in B. J. Best, ed. *The threatened forests of south-west Ecuador.* Leeds, U.K.: Biosphere Publications.

KESSLER, M. (in prep) Zur verbreitung, ökologie und bedrohung einiger sudwestekuatorianischer bromelien. Unpublished manuscript.

KIFF, L. (1991) unpublished notes on a Western Foundation of Vertebrate Zoology ornithological survey in south-west Ecuador.

KOEPCKE, H. W. (1961) Synökologische studien an der westseite der Peruanischen Anden. *Bonner Geogr. Abh.* 29: 1-320.

KRABBE, N. (1990) Red Data Book preliminary account for Grey-cheeked Parakeet *Brotogeris pyrrhopterus*. Unpublished.

KRABBE, N. (1991) *Avifauna of the temperate zone of the Ecuadorian Andes.* Copenhagen: Zoological Museum.

KUTZBACH, J. E. AND GUETTER, P. J. (1984) The sensitivity of Monsoon climates to orbital parameter changes for 9,000 years BP: experiments with the NCAR general circulation model. Pp. 801-820 in A. L. Berger, ed. *Milankovitch and climate.* Part 2. Amsterdam: D. Reidel.

KUTZBACH, J. E. AND GUETTER, P. J. (1986) The influence of changing orbital parameters and surface boundary conditions on climate simulations for the past 18,000 years. *J. of Atmospheric Sc.* 43: 1726-1759.

LAUER, W. (1988) Zum wandel der vegetationszonierung in den Lateinamerikanischen tropen seit dem höhepunkt der letzten eiszeit. Pp. 1-45 in H. J. Buchholz and G. Gerold, eds. *Lateinamerikaforschung.* Hannover: Jahrb. Geogr. Ges. Hannover.

LECK, C. F. (1979) Avian extinctions in an isolated tropical wet-forest reserve, Ecuador. *Auk* 96: 343-352.

LECK, C. F., ORTIZ-CRESPO F. I. AND WEBSTER, R. (1980) Las aves del Centro Cientific Río Palenque. *Separata de la revista de la Universidad Católica (Quito, Ecuador)* 27: 75-90.

Levêque, R. (1964) Notes on Ecuadorian birds. *Ibis* 106: 52-62.

Leyden, B. W. (1985) Late Quaternary aridity and Holocene moisture fluctuations in the Lake Valencia basin, Venezuela. *Ecology* 66: 1279-1295.

Luther, H. (1990) Bromeliad species recorded, Provinces El Oro, Azuay, Loja. Unpublished.

MacBryde, O. H. (1987) Machalilla National Park, Red Data Sheet preliminary draft. Unpublished.

de Macedo, H. (1978) Redécouverte du cracidé *Penelope albipennis* dans les forêts seches du nord-ouest du Perou. *C. R. Acad. Sci. Paris* 287 (Ser. D): 265-267.

de Macedo, H. (1979) Redescubrimiento de la Pava Aliblanca *Penelope albipennis* Taczanowski 1877. *Bol. de Lima* 1.

McColm, M., Neill, D., Werling R. A., Bolton, M. (1994) A proposal for the Jatun Sacha Bilsa Biological reserve. Unpublished.

McNab, A., Checker, M., Best. B. J. and Clarke, C.T. (1991) Aves en peligro de extinction en las montañas de Sozoranga. *Boletin informativo sobre biología, conservación y vida silvestre* (Loja, Ecuador) 2: 51-98.

Mace, G. M and Stuart, S. (1994) Draft IUCN Red List Categories. *Species* 21-22 : 13-24.

Marchant, S. (1958) The birds of the Santa Elena Peninsula, S.W. Ecuador. *Ibis* 100: 349-387.

Marchant, S. (1959) The breeding season in S.W. Ecuador. *Ibis* 101: 137-152.

Marchant, S. (1960) The breeding of some S.W. Ecuadorian birds. *Ibis* 102: 349-382; 584-599.

Markgraf, V. (1989) Paleoclimates in Central and South America since 18,000 BP based on pollen and lake-level records. *Quat. Sci. Rev.* 8: 1-24.

May, R. M. (1988) How many species are there on earth? *Science* 241: 1441-1449.

Meyer de Schauensee, R. (1970) *A guide to the birds of South America.* Philadelphia: Academy of Natural Sciences of Philadephia.

Mudd, A. (1991) Report on an environmental reconnaissance of the mountains of Chindul and Mache, northern Manabí and southern Esmeraldas Provinces, coastal Ecuador. Unpublished.

Mueller-Dombois, D. and Ellenberg, H. (1974) *Aims and methods of vegetation ecology.* New York: John Wiley and Sons.

Müller, P. (1973) *The dispersal centres of terrestrial vertebrates in the Neotropical realm.* The Hague: Dr. W. Junk.

Munday, M. and Munday, G. (1991) The vegetation and flora of sòuth-western Ecuador. Pp. 19-28 in B. J. Best and C. T. Clarke, eds. *The threatened birds of the Sozoranga region, south-west Ecuador.* Cambridge, U.K.: International Council for Bird Preservation (Study report 44).

Munday, M. and Munday, G. (1992) The climate of south-west Ecuador. Pp. 7-78 in B. J. Best, ed. *The threatened forests of south-west Ecuador.* Leeds,

U.K.: Biosphere Publications.

MYERS, N. (1988) Threatened biotas: "hotspots" in tropical forests. *Environmentalist* 8: 243-256.

NELSON, G. AND PLATNIK, N. I. (1981) *Systematics and biogeography: cladistics and vicariance*. New York: Columbia University Press.

OCHSENIUS, C. (1984) Late pleistocene aridity in the Neotropics as a cause of extinction of the South American land megafauna. *Zbl. Geol. Paläont.* Teil I. 11/12: 1691-1699.

PARKER, T. A. (1991a) On the use of tape recordings in avifaunal surveys. *Auk* 108: 443-444.

PARKER, T. A. (1991b) Unpublished notes on bird surveys in western Ecuador in 1991.

PARKER, T.A., SCHULENBERG, T. S. GRAVES, G. R. AND BRAUN, M.J. (1985). The avifauna of the Huancabamba region, northern Peru. Pp. 169-197 in P. A. Buckley, M. S. Foster, E. S. Morton, R. S. Ridgely and F. G. Buckley, eds. *Neotropical ornithology*. Washington D.C.: American Ornithologists' Union. (Orn. Monogr. 36).

PARKER, T. A., SCHULENBERG, T. S., KESSLER, M. AND WUST, W. (1995) Species limits, natural history and conservation of some endemic birds from north-west Peru. *Bird Conserv. Internatn.* 5.

PARKER, T. A. AND CARR, J. L., EDS. (1992) *Status of forest remnants in the Cordillera de la Costa and adjacent areas of southwestern Ecuador*. Washington D.C.: Conservation International. RAP Working Papers 2.

PAYNTER, R. A. (1972) Biology and evolution of the *Atlapetes schistaceus* species-group. *Bull. Mus. Comp. Zool.* 143: 297-320.

PAYNTER, R. A. AND TRAYLOR, M. A. (1977) *Ornithological gazetteer of Ecuador*. Cambridge, Mass. U.S.A.: Museum of Comparative Zoology.

PEARSON, O. (1982) Distribución de pequeños mamiferos en al altiplano y los desiertos del Peru. Pp. 263-279 in P. Salinas ed. *Actas del VIII Congreso Latinamericano de Zoologia Neotropical.*

PETERSON, R. T. AND CHALIF, E. L. (1973) *A field guide to Mexican birds*. Boston: Houghton Mifflin.

PRANCE, G. T. ED. (1982) *Biological diversification in the Tropics*. New York: Columbia University Press.

PRONAREG (1978) *Ecuador Mapa Bioclimatica* 1:1,000,000 and *Mapa Ecologico* 1:1,000,000. Quito, Ecuador.

PULIDO, V. (1991) *El Libro Rojo de la fauna silvestre del Peru.* Lima, Peru: Instituto Nacional de Invesigación Agraria y Agroindustrial/World Wildlife Fund-U.S./U.S. Fish and Wildlife Service.

RAUH, W. (1977) Bromelienstudien, 2. *Mitteilungen. Akad. der Wiss. Mainz, Nr.* 21: 21-28.

RAUH, W. (1984) Bromelienstudien, 17. *Mitteilungen. Akad. der Wiss. Mainz, Nr.*

8: 563-567.

RICHTER, M. (1991) Methoden der klimaindikation durch planzenmorphologische merkmale in den kordilleren der Neotropis. *Die Erde* 122: 267-289.

RIDGELY, R. S. R. (1981) The current distribution and status of mainland Neotropical parrots. Pp. 233-384 in R. F. Pasquier, ed. *Conservation of New World parrots*. Washington, D.C.: Smithsonian Institution Press for the International Council for Bird Preservation (Techn. Publ. 1).

RIDGELY, R. S. R. (1991a) unpublished report of Academy of Natural Sciences of Philadelphia ornithological surveys of the Machalilla National Park (Cerro San Sebastian), western Ecuador.

RIDGELY, R. S. R. (1991b) unpublished report on an Academy of Natural Sciences of Philadelphia ornithological survey of Manta Real, south-west Ecuador.

RIDGELY, R. S. R. AND GREENFIELD, P. J. (in prep.) *The birds of Ecuador.*

RIDGELY, R. S. R. AND ROBBINS, M. B. (1988) *Pyrrhura orcesi:* a new parakeet from south-western Ecuador, with systematic notes on the *P. melanura* complex. *Wilson Bull.* 100: 173-182.

RIDGELY, R. S. AND TUDOR, G. (1989) *The birds of South America, volume 1: the oscine passerines.* Oxford, U.K.: Oxford University Press.

RIND, D. AND PETEET, D. (1985) Terrestrial conditions at the last glacial maximum and CLIMAP sea-surface temperature estimates: are they consistent? *Quat. Res.* 24: 1-22.

ROBBINS, M. B. AND RIDGELY, R. S. (1990) The avifauna of an upper tropical cloud forest in south-western Ecuador. *Proc. Acad. Nat. Sci. Phil.* 142: 59-71.

ROSEN, D. E. (1978) Vicariant patterns and historical explanation in biogeography. *Syst. Zool.* 27: 159-188.

RZEDOWSKI, J. (1978) *Vegetación de Mexico.* Mexico: Editorial Limusa.

SALAZAR, A. P. AND HUBER, R. M. (1982) Ecuador's active conservation programme. *Parks* 6: 7-10.

SALO, J. (1987) Pleistocene forest refuges in the Amazon: evaluation of the biostratigraphical, lithostratigraphical and geomorphological data. *Ann. Zool. Fennici* 24: 203-211.

SARMIENTO, G. (1975) The dry plant formations of South America and their floristic connections. *J. Biogeogr.* 2: 233-251.

SCHULENBERG, T. S. AND T. A. PARKER, (1981). Status and distribution of some north-west Peruvian birds. *Condor* 83: 209-216.

SIMBERLOFF, D. S. (1986) Are we on the verge of a mass extinction in the tropical forests? Pp. 165-180 in D. K. Elliot, ed. *Dynamics of extinction.* New York: Wiley.

SIMPSON, B. B. (1983) An historical phytogeography of the high Andean flora *Rev. Chilena de Hist. Nat.* 56: 109-122.

SMITH, L. B. AND DOWNS, R. J. (1974-1979) *Bromeliaceae,* Vols. I-III. Flora Neotropica 14/1-3. New York: Hafner Press.

STADTMÜLLER, T. (1987) *Cloud forests in the humid tropics, a bibliographic review.* The United Nations University.

STATTERSFIELD, A. J., CROSBY, M. J., LONG, A. J. AND WEGE, D. C. (in prep.). *A global directory of Endemic Bird Areas.* Cambridge, U.K.: BirdLife International.

SUÁREZ DE FREITAS, G. (1990) *Diagnóstico del sistema peruano de áreas naturales protegidas y recomendaciones para su administratión.* Lima, Peru: Fundación Peruana para la Conservación de la Naturaleza.

SVENSON, H. K. (1946) Vegetation of the coast of Ecuador and Peru and its relation to the Galápagos Islands. *Amer. J. Bot.* 33: 394-498.

TERBORGH, J. (1974) Preservation of natural diversity: the problem of extinction prone species. *BioScience* 24: 715-721.

TERBORGH, J. AND WINTER, B. (1983) A method for siting parks and reserves with special reference to Colombia and Ecuador. *Biol. Cons.* 27: 45-58.

TERBORGH, J. (1986) Keystone plant resources in the tropical forests. Pp. 330-334 In M. E. Soulé, ed. *Conservation biology: the science of scarcity and diversity.* Sunderland, Massachussets: Sinauer Associates.

THIRGOOD, S. J. AND HEATH, M. F. (in press). Global patterns of endemism and the conservation of biodiversity. In C. J. Humphries and R. I. Vane-Wright, eds. *Systematics and conservation evaluation.* U.K.: Oxford University Press.

TOSI, J. A. (1960) *Zonas de vida natural en el Peru.* Inst. Interam. de Ciencies Agricolas de la OEA, Proyecto 39.

TRICARD, J. (1985) Evidence of Upper Pleistocene dry climates in northern South America. Pp.197-217 in I. Douglas and T. Spencer, eds. *Environmental change and tropical geomorphology.* London: Allen and Unwin.

TROLL, C. (1968) The cordilleras of the tropical Americas. Aspects of climatic phytogeographical and agrarian ecology. Pp. 15-56 in C. Troll, ed. *Geo-ecology of the mountainous regions of the tropical Americas.* Bonn: Colloquium Geographicum, Bd. 9.

VALVERDE, F. M. (1991) *Estado actual de la vegetación en la Cordillera Chongón-Colonche.* Ecuador: University of Guayaquil Inst. de Inv. de Rec. Nat., Fac. de Ciencias Nat.

VALVERDE, F. M., TAZAN, G. AND RISSO, C. (1979) *Cubierta vegetal de la peninsula de Santa Elena.* Guayaquil, Ecuador: University of Guayaquil Department of Botany publication No. 2: 1-236.

VANE-WRIGHT, R. I., HUMPHRIES, C. J. AND P. H. WILLIAMS (1991) What to protect? - systematics and the agony of choice. *Biol. Conserv.* 55: 235-254.

VARESCHI, V. (1980) *Vegetationsökologie der Tropen.* Stuttgart: Ulmer.

VAURIE, C. (1980) Taxonomy and geographical distribution of the furnariidae (Aves, Passeriformes). *Bull. Am. Mus. Nat. Hist.* 166: 1-357.

VUILLEUMIER, F. (1978) Remarques sur l'echantillonage d'une riche avifauna de l'Ecuador *Oiseau et R. F. O.* 48: 21-36.

VUILLEUMIER, F. AND SIMBERLOFF, D. (1980) Ecology vs. history as determinants of patchy and insular distributions in high Andean birds. *Evol. Biol.* 12: 235-379.

WANN, J. D. (1992) *Conservation in southern Ecuador: the relations between human pressures, forest vegetation and the conservation of north Andean endemic birds.* University of Dublin. (Unpublished MSc thesis).

WEBERBAUER, A. (1911) *Die planzenwelt der peruanischen Anden.* Veget. Erde. 12, Leipzig.

WEBERBAUER, A. (1945) *El mundo vegetal de los Andes Peruanos.* Lima: Ministerio de Agricultura.

WHITMORE, T. C. AND PRANCE, G. T. EDS. (1987) *Biogeography and Quaternary history in Tropical America.* Oxford: Clarendon Press.

WIEDENFELD, D. A., SCHULENBERG, T. S. AND ROBBINS, M. B. (1985) Birds of a tropical deciduous forest in extreme north-western Peru. Pp. 305-315 in P. A. Buckley, M. S. Foster, E. S. Morton, R. S. Ridgely and F. G. Buckley, eds. *Neotropical ornithology.* Washington, D.C.: American Ornithologists' Union. (Orn. Monogr. 36).

WIJMSTRA, T. A. AND VAN DER HAMMEN, T. (1966) Palynological data on the history of tropical savannas in northern South America. *Leidse Geol. Med.* 38: 71-90.

WILCOX, B. A. (1982) *In situ* conservation of genetic resources: determinants of minimum area requirements. Pp. 639-647 in J. A. McNeely and R. Miller, eds. *National parks, conservation and development: the role of protected areas in sustaining society.* Washington D.C.: Smithsonian Institution.

WILLIAMS, R. S. R. AND TOBIAS, J. A. (1994) *The conservation of southern Ecuador's threatened avifauna: final report of the Amaluza 1990-1991 projects.* Cambridge, U.K.: BirdLife International (Study Report no. 60).

WURDACK, J. J. (1980) *Melastomataceae.* Flora of Ecuador No.13. Göteborg, Sweden.

■